AROUND THE WORLD
WITH 1000 BIRDS

This book is dedicated to James, so that he will understand, and to Naaz, because she did.

AROUND THE WORLD WITH 1000 BIRDS

Russell Boyman

TravellersEye

Published by
TravellersEye Limited
Colemore Farm
Colemore Green
Bridgnorth
Shropshire
WV16 4ST
United Kingdom

British Library Cataloguing in Publication Data
A catalogue record for this book is available from the British Library.

ISBN: 1 903070 16 3

Printed and bound in Great Britain by Cox & Wyman, Reading, Berks.

Contents

Foreword
by Tony Soper

Russell Boyman gave himself a six-month challenge for a ticking marathon (a tickathon?) and in this book he tells all. It's an impressive undertaking.

I suppose I *might* have seen a thousand species in my life but as I've never kept a list I'll never know. But I once lectured on a splendid P&O cruise liner when my family spent three months sailing round the world. My seven-year-old son, Jack, found Cassin's Auklet in the swimming pool before breakfast. We watched Albatrosses power-glide across the wake. We sailed through a sea of Shearwaters inside the Great Barrier Reef and rescued some from a sticky end (becoming curry for the crew). While we stretched out on the games deck, doing our daily dozen, the Frigatebirds would soar overhead, doubtless sizing us up for dinner. I have the fondest memories of a truly disgusting rubbish dump outside Djibouti, where the gently rotting carcasses of donkeys and camels were attended by Abdim Storks, Ibises and an assortment of raptors. All this and superb palm-fringed beaches too.

One of my most enthusiastic fellow sailors was a lister. Together, we clocked up 650 species. Now any proper birder will tell you that the first few hundred are a doddle. It gets infinitely harder to top them up. Harder it might be, but without doubt a lot of fun. One of the best things about an interest in birds is that it provides the perfect excuse for travelling to distant parts. Russell Boyman took full advantage of his opportunity and now we get the pleasure of reading all about it.

Introduction

Any serious birders reading this book will know that trying to see 1000 birds in a six-month trip around the world is not especially impressive – the current world record in a single month, for example, stands at 1674, visiting eight countries. Certainly if *all* I had been trying to do was to see birds I could easily have doubled my eventual total.

But that's not *just* what this book is about. It's about visiting some of the great places for other wildlife, like whales, tigers and lions, not all of them good for birds by any means, and to find out what the rest of the world is like beyond my cosy, corporate, home-counties lifestyle. It's also about a journey of a different kind – the story of how a middle-aged man left his career, family and the girl he loved behind to discover not just birds or animals, but something about himself along the way.

This is a book for anyone who has ever wanted to see an Albatross or a Penguin in the wild, to stand next to a fluking whale or discover a tiger hidden in the long grass, but equally it is for anyone who is cutting the lawn or stuck on a commuter train and is thinking 'There has to be more to life than *this*'.

There is. Go out there and discover it. I did.

Russell Boyman
July 2002

Acknowledgements

I wanted to write a book for such a long time, without ever appreciating just how much hard work it can be. So my first thank you ought to be to those who held my hand during the six months it took to complete this book.

These include my editor Solveig Gardner Servian, whose encouragement and support was never less than total throughout, and Carol Lucas, who painstakingly copy-edited the book, and its Bird List, with a patience and attention to detail that puts all birders to shame.

However, the book would never had happened without the foresight of my publisher at TravellersEye, Dan Hiscocks, who had faith in the project despite my lack of experience. Thank you also to Debbie and Nick Sparey, and Charlie Hiscocks, for introducing me to Dan when this was just a pipe dream.

Gold stars are also due for Tony Soper, for writing the foreword when he really didn't have the time, and Richard Newton, whose own adventures abroad and kind advice gave the inspiration to write about mine. Cheers also to Alan Martin for his critique, and to Anthony Garvey for his researching skills.

Despite our obvious differences, without Rupert (he'll know who he is) I would never have had the money to do this trip, and I'll demonstrate here how far gratitude can go by saying thanks to him also. The Trip couldn't have happened without lots of careful planning, and the help from Quest Travel in Kingston, and all at Wildlife Worldwide, meant that the whole six months passed virtually without a hitch as far as travel arrangements are concerned.

No trip would be complete without the people who made it so memorable and, so, a mention in dispatches here for all my guides and drivers along the way, and a special thank you to Simon and Cate Bellamy in Sydney, who put me up (and up with me) in difficult circumstances, and Alison Ells, who was great company as we toured New Zealand together.

I couldn't have undertaken the Trip without the help of my brother Tim, who ran my finances whilst I was away, and who with his wife Jo looked after me when I got back to the UK with nowhere to live. Thanks also to my business partner, Colin Jelfs, who indulged me as I struggled to finish this book when I should have been doing my day job for our new company, Articulate Marketing.

Finally, a thank you to the two people to whom this book is dedicated. I hope my son, James, grows up to read this and discover a little more about his Dad than he thought he would. I would certainly never hold him back if he wanted to see the world as I did. And, of course, a heartfelt debt of gratitude to Naaz, who shared this adventure with me wherever in the world I was, and who was never far from my thoughts throughout. You gave me two fantastic years; I hope you find what you're looking for.

1 Mid-life Ibis

I watched that peculiarly English landscape, a green and auburn patchwork of fields and hedges and woods, disappear under the wing of the aircraft before it was lost for good in the low cloud. Well that's the last I'll see of that for a while, I thought.

I settled back as the plane continued to climb, closed my eyes and tried to encapsulate exactly what it was that had driven me to give up my job, my family and the girl I loved to go travelling around the world for six months in pursuit of birds, animals and other adventures, and at the ripe old age of forty-two.

There was no single answer to that question. Perhaps a realisation that there was a world beyond my desk at work, or the compulsion to seek out some of the globe's greatest wildlife sites before we managed to ruin them all, or even that this was a window of opportunity in my life that wouldn't come around again.

I thought ahead to the adventures I had been planning for the last three years. A journey through the Americas, across to Australasia, through Southeast Asia and then back via India and South Africa. Whales in Canada, Penguins in Argentina, tigers in India and lions in South Africa. The sunshine in Florida, the waterfalls in Brazil, my friends in Sydney and the women in Bangkok. A trip to conquer my wanderlust forever.

And, of course, there was my mental checklist, my mantra, my muse throughout – the race to see a 1000 bird species on my way around the world. All in all, it promised an awful lot of living ahead.

It had to. Whilst I was busy indulging myself and fulfilling a long-held ambition, life would be going on without me. My son James would be growing taller by the day, improving his football skills, coping with a new school year. Meanwhile the advertising industry, which had nurtured me since *my* schooldays, would carry on putting noses to grindstones and cutting throats in the best of adland traditions. God knows if it would have space for me when

I got back, or whether I would want it if it did.

And, the girl I loved would be preparing to emigrate to Switzerland. I didn't know as I sat there on the jumbo jet out of London whether I would ever see her again.

But, there was no looking back now. My mood improved as we rose above the clouds to that world of perpetual sunshine in which all air travellers live. I looked across at the field of woolly white clouds that stretched over the Atlantic below. You owe it to yourself to go for it, old son, I said to myself. Now go out and enjoy it. The real world can wait.

I can still remember the exact moment I told the boss he could stick the job.

We – that is the media buying company I had helped launch – had just been fired by our biggest client, following a three-month competitive pitch that had been fraught with politics and unbelievable pressures from head office.

In the same month I had finally got a very messy divorce from my wife of some thirteen years (unlucky for some, I remember thinking). To say I was not in a positive frame of mind was an understatement.

Rupert, my MD, emerged from the bar carrying two pints of cider, a surreal enough sight on an already unusual afternoon, since we had all assumed that he had forgotten where his pockets were years ago, let alone his wallet. He put down mine with the satisfied grin of a man who only bought drinks on special occasions – birthdays, weddings, Christmas and on days when the future of his company was critically threatened.

I took one sip, briefly looked at the hazy afternoon sun setting on a typically humid and sweaty Central London, took a deep breath and decided to simply say what I thought.

'Rupe,' I said with a studied grace I didn't really feel, 'you know this really is a bloody awful job. You can keep it. I mean that.'

To his credit he agreed, up to a point. No one had worked harder than he to keep the client happy, but finally the unthinkable had happened and he had watched as a third of the company's income walked out the door and took residence with a competitor – one that had offices opposite ours, to add insult to injury.

But that benign grimace that passed for gallows humour from Rupert barely disguised what he was thinking – something along the lines of 'Yes, Russell, I'm fed up too, but tomorrow we all have to go to work to get that income back, don't we, hmmmmm?'

Except that for the first time I didn't feel like that. I genuinely felt that now was the perfect never to be repeated moment to break out of the corporate mould and take a risk for the first time in my life. I didn't have a dream as much as wanted to avoid the nightmare of another year promising clients the earth whilst they paid us less for the privilege, and the constant stress and hassle that came with it.

Not that this was unusual in the media and advertising industry; in fact it passed as normal behaviour – it was business as usual. But it was a business that I had grown so very, very bored with, and one in which my own specialisation had somehow turned into a commodity without me even noticing.

I couldn't make a difference any longer, and it had taken the loss of our biggest client to ram home how spiritually redundant the whole thing had become for me.

Working closely with Rupert hadn't helped. We had worked together for ten years without really knowing each other. In fact the only time I had ever been to his house (a huge mansion in a fashionable London square) had been a couple of weeks previously, when he called me back from holiday to work on the pitch. In the media buying business this was not unusual, as in a pitch it was always all hands to the pumps – but on a *Saturday night?*

Rupert had employed me in unusual circumstances anyway. Although I had made the shortlist for the job, he had in the end chosen a good friend of mine, someone I had helped train, ahead of me. He had accepted, and Rupert had trumpeted all over the trade press how he had landed 'the best negotiator in London'. Except that this chap had then changed his mind, that is, accepted a hefty pay rise from his current company, and I had hurriedly been offered the job instead.

I accepted, without truly appreciating that I was to forever be saddled with the epithet of 'the *second best* negotiator in London'. From that point on, I felt whatever I did was never good enough for good old Rupe.

We were very different of course. I played the advertising industry game for all I was worth – work hard but then play hard, never turn down an invitation to lunch or to some corporate hospitality, because in the end you might be working to midnight the next night or all the following weekend.

To Rupert, work was a science, and one to which he could apply his vast logical brain and be good at. He had few weaknesses – as he constantly reminded us all – except for a complete lack of humility and charm.

To be fair, these were two of the few skills I possessed. I had prospered for over twenty years in the only profession I had ever had, through a mixture of hard work, luck, personality and the ability to strike a deal. I had risen to Managing Partner of a top ten company in its field, was reasonably well paid, relatively popular (for a boss), but hadn't learnt anything new in years.

Rupert thought of me as a jumped up barrow boy, a necessary evil once upon a time but whose time was passing – funny, popular, but a growing liability in a world where margins were tight and everybody had to try harder tomorrow than they did today. He hated the casual way I ran my department, whilst being jealous of the fact that I had 'the common touch' as he put it.

All of which was fine by me. I regarded Rupert as an intellectual snob, with a superiority complex of biblical proportions unless you had attended the London School of Economics (like him) or at least Oxbridge. I had quit school after my A levels, and opted for the lure of the salary cheque rather than university, and he could never understand how someone who was obviously not stupid hadn't done more with his education.

But then that was him all over – the son of two magistrates, who had learnt to be judgemental about everyone at a very early age.

I knew I didn't want to work at the company any longer, but I didn't know what it was that I *did* I want to do. I just knew that it had to be something radical.

For many years, whilst I had been the corporate slave and faithful husband, the word 'radical' would not have been in my repertoire. I had found my niche, and was busy exploiting it for all I was worth. Keep your head down and just beat the next bastard. But two things had happened to me that had changed all that. First, my marriage finally imploded and, secondly, I got lucky with company profitability.

Life with my ex-wife Sue had been great for some years after we got married, but the stress of two miscarriages had opened big cracks in our relationship, and these were opened further and fractured by a lifestyle that forced us further apart. For me, play had to follow work, and a couple of nights a week I would go out with the people from the office or to an industry function.

There would always be lots to drink, and although I never did drugs it was obvious that there was plenty of this around as well. I was less successful, though, at ignoring the seemingly unending array of young, sexy and available girls that seemed attracted to the advertising industry like television viewers to *Coronation Street*.

I had never been particularly successful with the opposite sex, but as I got older (and, tellingly perhaps, more powerful in the industry) I found for the first time that these women were not just the strict preserve of that good-looking young lad in accounts. I was never exactly a Lothario, but had my head turned enough to realise that I had changed.

Having met Sue whilst very young, in my early thirties I was suddenly somehow reliving those lost years, but with more success and significantly less acne. And, I couldn't do that *and* be married.

At the weekend, time was divided between my son James, running our expensive (*too* expensive as it turned out) new house and indulging my growing obsession – to watch birds. How prophetic that it was my burgeoning interest in ornithology that helped end my marriage, which in turned help me quit the job and further indulge my passion.

I firmly believe that all men have the hunter-gatherer instinct in them, which naturally leads them to collect things throughout their life. As a child this could be football cards or, in my day, stamps and coins. Later, it might be girlfriends, or motorbikes. By the time middle age beckons, a golf handicap is the more likely culprit.

Well, I only had a few girlfriends, never rode a motorbike and detest golf to this day, but I had always been interested in nature and had been fascinated by birds in my early teens. There was something about their beauty and unobtainability that touched a chord in me.

But it was not until Sue and I moved in together in 1982, and she became unwell with a series of illnesses, that I found myself visiting her in hospital and having lots of spare time to myself. I used this to go walking on Wimbledon Common, initially to walk off my worries over Sue, but later to explore every nook and cranny in an effort to increase my bird list for the area.

The operative word here is *list*. The seeing and identification of the bird was only part of it, the *recording of it* was the truly anal yet compulsive part to us hunter-gatherers. No wonder 90 per cent of birders are men.

Although Sue eventually emerged from hospital, by this time it was me who had caught the bug. Soon I was helping the local recorder build his official list, had started travelling further afield to see more birds, went on organised trips with bird clubs and spent a huge sum on my first decent telescope. I was hooked.

No one in my world of advertising could understand my particular obsession. They were mostly still into girls, motorbikes and golf anyhow. I lost count of the people who said to me, 'Russell, we know which kind of birds *you're* interested in'.

Ha ha! The truth was I was more interested in chasing birds than I ever was in . . . well, chasing . . . birds. I never went anywhere in the car without a pair of binoculars to hand, and started to fantasise about beautiful telescopes made of steel and chrome. It quickly became an obsession.

By the time my marriage was on the rocks, some ten years later, I had graduated to a week's separate holiday every year, specifically to go bird-watching. Indeed our last holiday abroad with James was to Majorca, especially chosen so I could go birding half the time and spend the other half on the beach with the family. And since I earned enough to always go on at least two good holidays a year, and provide all mod cons for the new house, I never once felt guilty.

When finally our marriage ended, we tried living together for James' sake, but it was an awful nightmare and in the end it got so bad that I moved in with my parents – a grown man sleeping in the same bed he had when he was sixteen at his parents house.

Bless them. They were more understanding than I ever thought possible.

We were officially separated that summer, and after a nightmarish legal process that neither of us could afford, divorced the following year, right in the middle of the Big Client Pitch. It was the worst week of my life.

Although the divorce itself was a sort of a score draw, the crippling mortgage payments meant that I had just about enough to live on – live being a relative term, of course. In my case, literally a relative term, as I had no prospect of moving out of my parent's house for some time. So I had little to give up in the material sense once I decided to pack up and go around the world.

The other thing that made it easier to give up the job when the crunch came was money – the security to take a risk knowing you could pay the bills after all.

Rupert, to his credit, knew how to run a financial spreadsheet, and had set up a deal whereby all the partners received a share of the profits, subject to us staying in place at the company for a set period of time.

This was originally designed to be the icing on the cake, to lock in the partners, but against all predictions Rupert's parsimonious ways bore unexpectedly juicy fruit. That icing wasn't just inches thick, suddenly it was several feet thick and none of us could believe our luck.

The quick-witted amongst you, however, would have seen the fatal flaw in my plan. Here I was, bored rigid with my job, living in one room at my parents place, newly divorced, seeing my son (if I was lucky) once a week, and soon to have the means to get out and do something positive and new with my life. The only snag was I had to work for another three years before I got a penny.

Well, I may not have gone to the LSE, but even I knew that I couldn't afford to leave until the end of the term of the profit share, which was worthless until the day it matured.

So, I was faced with three more years of pushing myself reluctantly into client meetings, annual negotiations with the television stations for airtime rates, tedious board meetings, untold new business pitches, not to mention endless financial squabbles with my ex-wife and blood-curdling conversations with my vulture of a solicitor, or the nightly trek back to my old bedroom in Essex at my parents house.

But I already knew, as I finished my cider outside the pub and watched the weary workers tramp home through the sticky end to the day, what was needed if I was going to survive those three years without going mad. I needed to know what all the agony was really for, to work out what I wanted to do, not just with the money but also with the second half of my life. I needed something to look forward to, something to help me forget both the past and the present.

What I needed was *a plan*.

You might think that having three years to plan the adventure of a lifetime was a good thing. True, it gave me plenty of time for reflection, but I found that planning what to do next was actually the easy bit; not changing my mind was the real challenge. I knew that walking out on a twenty-five-year career could be the worst thing I'd ever done, and that getting back in was not going to be easy.

But I knew deep down that I'd reached a crossroads in my life, needed a new challenge and a rest from the comfort blanket of the only job I'd ever known. If I went back now, I would never forgive myself.

I said to myself that kids were flexible, almost immune to the passage of time, and that if I were to go off and do something new, perhaps somewhere else, James would not even notice after a while – anyway, he would be happier if his Dad was more content and fulfilled. And as for Sue, well, the bloody job could disappear at any moment anyhow, better to make a planned departure that would set me up for the rest of my career, so I could continue paying those bills. 'Shit or bust' was my motto. I'd given up all the things that I used to value – house, car, possessions – now it was time to revaluate what I thought was really important. Or, as George Michael once sang it, 'Now I'm gonna get *myself* happy'. Freedom indeed!

So it was settled, the money would buy me the chance to take some time off. I would have to fund my payments to Sue and James whilst I was unemployed, and had to budget also for the inevitable time spent sitting on my backside when eventually I did decide to start work again. And finding out exactly what I wanted to do when that time arose was one of the things about myself I wanted to discover.

A year sounded a bit like too much of a risk. So, six months it was then. It must be everyone's dream to plan for a six-month holiday but, believe me, doing it for real is a daunting task. You didn't want to spread yourself too thinly, but equally wanted to cover all your must-see places. So what would be the focus, the theme of the Trip (as I had now started to refer to it)?

Nobody who knows me will be surprised to learn that the answer to this was '*birds*'!

Here was a fantastic opportunity to see some of the great birding sites. I had spent years locked in my office, chained to the desk whilst listening to Birdline, a telephone service that updated details on rare birds arriving in the UK. I always missed them, unless by some miracle they happened still to be around on a Sunday, or I could hurriedly take a day off.

Just think, if I was not working I could go birding every day for six months. My UK species list would be up to the magical 400 level in no time (this being the threshold that needs to be passed by any UK birder worth his or her salt). Except that the next word that came to me after 'birds' was '*travel*'!

I had always had a deep-rooted desire to travel ever since I was old enough to go on holiday without my parents. I was fascinated by the whole deal, not just the usual things like the weather and the scenery, but all the mundane stuff about foreigners' lives – the currency, the time difference, the television stations, the food. I'd find myself occupying my thirty-minute walk to Waterloo station deep in thought constructing a list (inevitably) of what men my age would eat in various countries around the world on any given day, and comparing it to my diet. Sad? Well yes, but it fulfilled both my desire to collect things and my intense curiosity about how the rest of the world lived.

But here it was, another opportunity to see the world, which had come along when I least expected it. No matter that you were supposed to do all this at twenty-two, not forty-two, I wasn't going to pass up the chance a second time around. No, here it was, my last chance to squash all those demons and make up for lost time.

So there it was – birds and travel – and, who knows, as a single man, perhaps some women along the way as well.

I strived to remember all those wildlife documentaries I'd seen and thought 'Boy, one day I'm going to go *there*'. All those conversations I'd had on my birding travels, where I'd said how good I thought, say, Majorca was for birds only for someone else to trump me and say 'Yes there were *thousands* of those in India last year'.

I disappeared under a mountain of wild life brochures. I knew there was already a danger of trying to cram too much into the six months I had, so first it was a question of editing. I looked into flight schedules and weather maps, and read up on bird migration patterns. Suddenly the Internet was a lifeline for me, and I began to spend large chunks of time in the office when I should have been worrying over sales figures looking into the optimum time to see American warblers in the South American rainforest instead.

All sorts of pipe dreams were discarded on the basis of cost or logistics. I even had to admit defeat over Rio de Janeiro, especially galling since I had dreamed of visiting Brazil since I had been captivated as a boy by the great Brazilian World Cup winning side of 1970.

Later in life I often said that my idea of heaven was to go to Rio, play beach football on the Copacabana and gaze up at the statue of Christ on Sugarloaf Mountain whilst having some teenage lovely, dressed only in a thong, massage my aches and pains away afterwards.

This noble dream came crashing down around my ears when I later heard lurid tales of bodyguards, muggings, street gangs, poverty, pollution, rip-off merchants and gunfights. No footy or massage was worth getting killed over and, in any case, there appeared to be no eco-tourism industry to speak of – unsurprising, since the government seemed much keener on turning their immense rainforest into cattle ranches and oilfields as quickly as possible.

Slowly, I pieced together a plan of action. Cape May in New Jersey in the USA was a must, the tales of bird migration there as huge populations of birds from North America moved south for the winter were legendary. By leaving the UK in October I would hopefully hit the sweet spot for migration there.

Another obvious venue was the Galapagos Islands, a dream destination for any birder. This was where Darwin formed his later theories on evolution, and for that reason alone was sacred to anyone with an interest in wildlife. Also, much of the wildlife was to be found nowhere else on earth, so remote and unique was the area.

I had loved the time I had spent birding in the rainforest in the neo-tropics on previous holidays, where the sheer profusion and beauty of many of the birds was almost unbelievable. As I was going to be in Ecuador anyway, it would be a shame not to explore the Amazon basin as well.

The other birding mecca that I could not afford to miss was in Rajasthan in India, and was called Bharatpur. This was a huge wetland reserve, which attracted thousands of birds wintering south of their breeding grounds in Siberia and Central Asia, including some particularly rare species that were the stuff of birding legend.

So, a bevy of birding was assured. I then began to look into other wildlife hotspots. Whales had been a passion of mine for over ten years, since Sue and I had spent an unforgettable two weeks amongst the grey whale breeding grounds in Baja, Mexico. Once having seen a whale, the humbling experience of sharing the ocean with one of the most gigantic yet gentle and intelligent creatures is an experience that stays with you for ever – guaranteed to raise the hairs on the back of your neck whenever the story is recounted.

I was very excited about seeing more whales, to add to the half a dozen or so species I had seen already. Vancouver Island I knew was renowned for orcas, and so Canada was fitted on to the front of the trip before the weather got too bad there (and how wrong was I about that?).

Patagonia in Argentina was famous for the ability to get close to the right whale, once the commonest whale on earth, now one of the rarest. A logical place to go after Ecuador, I thought.

Kaikoura in New Zealand was another famous destination for whale-watching, where the continental shelf brought sperm whales, usually a deep-sea species, close in shore.

As I was going to be in India anyway, I promised myself that I would find a tiger. I knew this was not going to be easy, so a number of reserves would need to be covered if I was going to stand any chance of seeing one.

Lastly, Africa was the obvious place to finish the trip. I had always fancied going to Cape Town, and so I decided to do my African safari in the Kruger National Park in South Africa.

I decided early on that there were activities on my wish list other than just watching birds and animals. Once I was in the USA, I planned to visit the Gettysburg battlefield in Pennsylvania, military history being another passion. Also, I wanted to visit the glitzy charms of South Beach in Miami, where I figured I could chill out for a bit and see some of the rest of Florida before the hard travelling began.

I had also been lucky enough to see some of the great waterfalls on the planet on previous holidays, first Niagara and then Victoria. I was in awe of the power and the beauty of these natural phenomena, and had made a mental note to try to see some of the others at some point.

Well, as I was going to Argentina anyhow, why not fly north and see the Iguaçu Falls, on the border with Brazil? I might even get across the border, although it was doubtful whether I could see the Sugarloaf from that far away, never mind the beach.

Birding was by nature largely a solitary pastime. But I must admit that the thought of spending Christmas and New Year alone did not fill me with joy, especially as I was likely to be missing James a lot at this point. Therefore, a visit to friends in Sydney seemed ideal. The city was probably my favourite place on earth, I hadn't visited it for ten years and it would be a great opportunity to kick back for a few weeks and celebrate with people I knew and liked. And, of course, New Zealand and its whales were right next door.

This just left a gap in my plan of how to bridge the gap between Australasia and India. I had about a month to play with.

I had really enjoyed my only other visit to Southeast Asia – the colour, the food, the temples, the culture – and the chance to find out more about this fascinating region was a must. I decided to fly into Bangkok, spend some time in Thailand and then move on to one other country, rather than spread my time too thinly. I didn't want it to be either too developed or too remote – a new experience, but one I didn't have to share with raving backpackers half my age. With my interest in warfare, Vietnam seemed the obvious choice – another continent would certainly help the bird list too.

The whole thing came to a total of twenty-four glorious weeks, nearly half a year. No less than a week anywhere, with nice long stays mixed up with shorter ones. Intensive, sure, but not heart attack material. Little did I know at this point that two other countries would unexpectedly appear on the itinerary before the trip was finished.

Also, the ten original countries felt like a weighty project, a worthwhile risk, and it really did go *all* the way around the world. There were no significant omissions, I felt, apart from Europe which I could visit anytime.

Putting the plan into action was relatively easy once I had the itinerary complete in my head. I went for mixture of good hotels and backpacker tours, reasoning that a mixture of experiences was a good thing and that, however close to the culture I wanted to get, I was occasionally going to value a good hot shower and a laundry service.

I also left lots of free time in countries where transport and infrastructure allowed me to bird on my own, but wanted to book arranged tours from the UK for areas where language or logistics might prove to be problems.

This was where a company I had used before, Wildlife Worldwide, came into their own. I took three of the tours from the brochure and asked them to customise them to suit my dates and requirements. This took all the hassle out of moving around in Ecuador, Argentina and India at the start of the process and made it someone else's problem, giving me maximum time birding in the field, in countries with lots of list potential.

Two months before the magic date of April 2000, I had finally committed to that plan, to the extent that I had paid my deposits even before I had officially resigned. There was no going back now.

There still seemed to be something missing however. Where was the vision, the mission in what I was doing? What could I tell people about it? Was it just a glorified holiday, or something more profound?

Then I decided that birds would be my goal, my trip's spirit level. They had been my original inspiration, and although I had adapted the Trip way beyond what would have delivered maximum numbers of birds, the hunter-gatherer in me told me I needed some type of challenge, a list, a target to make it even more personal for me.

That's when *Around the World with 1000 Birds* was originally conceived.

In itself, the total of 1000 birds would have been a relatively easy target in six months across four continents to any of the really serious birders I had met. The world record life list stood at something over 8000, and it was possible to get close to 1000 just by spending enough time solely in South America, if you had the time, knew where to go and that's all you wanted to do.

Except that I had to acknowledge that I wasn't that single-minded. This was a whistle-stop tour, with birds only being a part of my adventure now. How many birds would I see from a beach in Sydney or a temple in Thailand?

At the time my own life list was around 1200, the product of some eighteen years' increasingly serious birding. I decided that seeing 1000 in total in just six months was a worthy target, and one I hoped I would not have to compromise the Trip too much to achieve.

It wasn't a grand total, but it was *my* total. And that was what this trip was all about: after many years of working for other people, of shelving all my dreams, this was six months for *me*, and me alone.

Of course, I knew that I ran the risk that I could have a great time and then ruin it if I didn't achieve the magical 1000 figure, but it was that challenge that I wanted to spur me on if time ran out and I was still short of the total.

And so, finally, I formally resigned to no great fanfare, began to explain why I was doing such a damn fool thing to friends and family, and prepared to spend the last six months at work winding down, seeing friends for the last time in a while and fine-tuning arrangements for the Trip.

However, nothing is ever that straightforward. If anyone can cock it up, you can rely on me to do it. Just when I thought I'd accounted for everything, something happened to me completely out of the blue. That's when I fell in love again.

And to think that the woman who nearly made me change my mind I didn't even fancy at first.

I first met Naazlin just two months before I was to resign and eight months before I planned to leave. She insisted that she had to come in and meet with me as the company she worked for was about to become our new recruitment consultants. She sounded frightfully posh but rather sweet; she flirted as much with me as I did with her and in the end I agreed, more out of curiosity than anything else. I'd long been fascinated by Indian women, and despite the cut-glass accent her name gave that much away at least.

'If she's as pretty as she sounds this will be fun anyway,' I thought, ever the sucker for a pretty face. Except that I didn't think she was pretty – at first. I met her in reception, and apart from being very short and very dark, there were two very striking things about her.

First, and at the risk of sounding corny, she had (and still does have) the biggest eyes I'd ever seen on a woman – huge, they seemed to fill half her face, were a deep hazel in colour and were framed by lashes so luxuriously long that they seemed absurdly false at first. Secondly, she had spirit, and I had always liked a woman with balls (preferably mine).

She perched on the edge of my sofa, as relaxed as if she had met me, a managing partner of her company's newest client, a thousand times before. She bamboozled me with a multitude of questions, insisted on meeting my staff and generally acted, in the nicest possible way, as if it was *her* office all along. Only later did I learn that she had been in the job just over a month.

Naaz, ever the keen one to impress, was on the phone daily and quite quickly we started to gel, in a professional way. She had the habit of calling on my private line and introducing herself as 'Hi, it's me'. The implied intimacy of this from the start began to fascinate me.

I decided to flirt with her unmercifully, as I was convinced she was the proverbial 'Asian Princess' and couldn't possibly be so self-assured once I started to work on her composure.

For Naaz's part, she had heard from many of the old-stagers in her office how I was an incorrigible flirt, and worse from others that had been around for a bit longer. She was determined not to let me embarrass her and not to let me get close to her, as I had with others over many a would-be telephone romance – as one girl once told me 'Russell, you know you give really good *phone*'.

Quickly – far too quickly, it seemed – we formed some kind of bond. We discovered our birthdays were a day apart, or more correctly fourteen years and one day as Naaz was only two-thirds my age – something that merely increased her appeal as far as I was concerned, especially as this did not seem to put her off.

Our calls became regular affairs, and soon became marathon sessions where we exchanged confidences. She told me all about her Swiss boyfriend; she seemed very loved up and they were enjoying themselves immensely on his periodic visits to London, or on hers to Zurich. I was beginning to realise that perhaps she wasn't the orthodox, well-behaved Muslim I'd thought she was.

Soon Naaz had talked me into buying her lunch at Bank, a new and expensive London eaterie, and when I had the temerity to cancel and rearrange, you'd think I had committed the ultimate sin. I wasn't going to cancel a second time, that was for sure. 'Boy, this girl is *hungry*' I thought.

If there was any doubt as to whether I was attracted to Naazlin or not, the lunch date proved it. She looked very different to our first meeting. She wore a pin-stripe suit, with a very short skirt (despite the weather) and underneath a tight lurex top that did nothing to hide her obvious charms. 'How did I miss those?' I remember thinking.

Over lunch, we clicked immediately. I was mesmerised by how she seemed so comfortable, half-undressed though she was, and yet hated the fact that the men on surrounding tables leered at her throughout. It was the first of many enigmas about Naazlin.

We both sensed I think that we had a kind of unique chemistry and, when she accepted an invitation to meet me for a drink one evening in Richmond, even I began to believe that this might just be going somewhere.

That night I told her all my secrets. I told her I had resigned, I told her about my marriage, I told her about my love life since I had left Sue. I seemed to commit part of my soul to her that night.

For her part, Naaz listened intently, and told me of her love for JJ her boyfriend but she too felt something click I think. She suggested a walk by the river, and as we stood there in the moonlight, watching the river, feeling the spring air cool against our faces, we embraced for the first time. From that point on I was in love with her, and only a couple of months after we had both met.

Our relationship moved at an incredible rate. We knew it was more than a friendship, but was certainly not a love affair. Naaz would let me get so close and then make it obvious that her body, at least, belonged to JJ. Enigmatic as always, this did not stop her showing most of it off with a series of very revealing outfits. I was so excited by her, I was even content to look and not touch.

We quickly began to see more of each other out of work. She invited me round to her tiny, overcrowded flat in Feltham, and cooked a delicious vegetarian curry for me. She took off my shirt, and gave me a therapeutic massage. She asked me to sleep over, changed into the skimpiest negligee but never once gave me the idea we would do anything other than actually *sleep* together.

Naaz was a delicious paradox. Her values were from the East, but she dressed like any career girl from the West, and constantly challenged the views of her family who were much more orthodox.

She was fiercely loyal to JJ, yet managed to dress provocatively without it ever seeming sluttish – she simply had a good body and a well-meaning exhibitionist streak. She always looked at me like I was the only man alive, and then felt so guilty afterwards. She had half a dozen other men she was friendly with, most of them ex-boyfriends. She collected men in the way I collected birds. All of them remained in love with her, and three of them proposed marriage to her in the six months before I left. I was getting used to such things happening when Naaz was around.

The fact that I was going away in a few months seemed to be our fail-safe. We knew that we were involved in something that had a momentum all its own and that was moving at an incredible speed, but convinced ourselves that it would not be destructive to her and JJ as it had to end in the autumn when I left . . . didn't it?

As spring turned to summer, we spent every moment together. It was so easy to be happy with each other, and we seemed to guess what each other was thinking half the time. She met James and got along famously with him. I told her I loved her, and could clearly see she loved me too but could not afford in the circumstances to actually tell me (although eventually she gave in). JJ knew everything and was content to give this new girlfriend of his as much space as she needed.

We had just returned from a riotous weekend together in Le Touquet when she told me that she felt that she and JJ had to be in the same country if they were to make any progress together, and that she was planning to move to Switzerland the following spring.

I was gutted, of course, but how could I object? I was leaving the country myself, and couldn't make any commitments to her of any kind beyond October. Plus, I could see it made sense for them – he had been very tolerant of the time we spent together, and she didn't want to betray that trust by not seeing how far they could go together.

I made up my mind that here was an incredible woman, one I was unlikely ever to have, but equally one I could never replace. I had the money to spoil her, and now the perfect excuse too – I may never see her again.

Our last couple of months together became a debauched festival of excess. We stayed in the best hotels in London after a night on the tiles. We spent a romantic weekend in New York at the drop of a hat, eating in the Windows on the World atop the World Trade Centre almost a year to the day before the tragic terrorist attack. We stayed in the Arts Hotel in Barcelona on another trip, misbehaving like two naughty schoolkids and daring each other to be more outrageous. We simply didn't care, it was if we were on borrowed time.

I even committed the cardinal sin for any birdwatcher. I was spending two weeks on the Scilly Isles, a must in October for any self-respecting birder as weather conditions usually meant something exotic would turn up, sometimes even from across the

Atlantic. The entire birding cognoscenti were usually present during this month, and you were nobody in birding circles if you were not there.

Well, I was there all right, but my mind was elsewhere. I remember standing in a rainstorm in Old Town Churchyard on St Mary's, trying to flush a Red Eyed Vireo that was understandably reluctant to brave the wet and show for the assembled throng of soggy anoraks. Most others were anxiously looking at their pagers for more information from around the islands, or on their CB radios giving updates to others. Meanwhile, I was on the mobile to Naaz in London, telling her how much I missed her.

It got so bad that I decided to interrupt the trip and come home to spend a weekend with her in Feltham. Only a mere 400 miles, £150 and eight hours travelling, and then all that again in reverse three days later. But I never regretted a thing (and also didn't miss any worthwhile birds, luckily.)

On the basis of 'nothing ventured, nothing gained', I proposed one wet afternoon, sitting in a pizza restaurant in Little Italy in New York, and gave her a diamond bracelet I had had made for her.

She was flattered, shocked and deeply troubled. I could see all the emotions running across her face. She loved me, but how could she betray JJ's trust any more than she had already? I was also about to disappear for six months. And, I was the wrong age, the wrong colour, the wrong religion and the wrong marital status to be accepted by her family. She never actually turned me down as such, but after a long tearful afternoon chatting about it, it was obvious to us both that it was not going to happen. I guess I never really expected it to; it was just something I knew I needed to do before I left.

I told Naaz that I loved her so much, that if she were to marry me I would do the unthinkable and cancel the Trip. I wasn't sure if I meant this, but in that heightened emotional state, I felt right then as if I did.

Luckily, Naaz would not test my resolve. There were bigger reasons why we could never be married. I'm not sure if I was more relieved than upset, but from that point on, in the last few weeks before my departure, we decided to squeeze every last ounce of love and fun from our time together – for who knew what lay over the horizon when I got back?

So, I was going around the world after all. About time I said my goodbyes then, I thought.

Saying goodbye to James was always going to be tough, although with the innocence of childhood he never judged me, or seemed to appreciate that I would be gone for some time, missing both Christmas and his birthday. At least I bought him a brand new computer, so we could email each other whilst I was away.

My parents' main emotion seemed to be one of dread. They sort of understood why I needed to do it, although the phrase 'more money than sense' cropped up more than once, and my Dad couldn't really reconcile why I couldn't struggle through life in a job I hated, as most of his generation had had to. I promised faithfully to keep in touch, and tried to avoid the multitude of questions about how I would cope when I got back. I didn't want to think about getting back, or about work at all, and didn't need another reason not to go. At least my brother Tim, equally fed up with the rat race, told me I was doing the right thing.

Around the advertising industry, my friends and colleagues greeted my news with a mixture of envy (that I had concocted a plan of escape and could afford it) and puzzlement ('what, you're going away, alone, to watch bloody birds?').

I had the requisite leaving do, of course, a comparatively low-key affair with everyone from the office and a few mates from around the business, but I was going through the motions and couldn't wait to get home with Naaz afterwards.

When the final day at work arrived, I was numb, not able to quite comprehend that I would never be back. Rupert gave what he thought was a humorous goodbye speech, describing me as the company's 'Minister of Fun' and a 'serial shagger' (I'd actually only ever slept with one of the staff, and that years before. This was positively homosexual behaviour by advertising standards).

After ten years' service, it was a predictably miserly goodbye. I went back to my desk, wrote an all-staff email saying that I thought I'd be remembered for many more achievements than just the ones Rupert attributed to me, and wished everyone good luck. A quick drink in the local and I was gone – ten years of work, all forgotten in a flash. Which just left Naazlin to say goodbye to.

In fact, we'd been saying goodbye to each other since New York. I wasn't to know at that stage that our relationship had many twists and turns still to come.

Of course, I cocked it up on the final day. I met her in our favourite wine bar, with all my suitcases and we had a tearful and solemn last meal together. I had planned a brave speech, but in the end I somehow lost the nerve. As I was trying to think of what to say, a cab pulled up at the kerb. Without even thinking, I put my bags in the front, gave Naaz what I hoped was a passionate kiss and jumped in the back.

As it sped away to Heathrow, I looked out the back window and saw this little, dark figure wandering forlornly across the road. I hadn't begun to say what I felt to her, and now it was too late.

I tried to make up for it on the mobile of course, but nothing could wind back time and let me start again. I couldn't believe that that was how I would see her in my mind every night for the next six months. I made up my mind then and there to find a way of saying goodbye properly.

I sat in the departure lounge and tried to convince myself everything would be OK, but I was upset that I had bungled the most important goodbye of my life and annoyed that I would have to carry this baggage around with me throughout what was supposed to be the adventure of a lifetime.

I sat down with my journal and wrote the first words of what was to become the template for this book. 'Why can't I ever do anything right?' it said. It could have been a metaphor for my life, the reason I was running away in the first place.

But running away I was. The Trip, three years in the planning, had finally begun.

2 Riders on the Storm:
Canada, October

I didn't really know what to expect from Vancouver. Well, it was near to Vancouver Island (obviously!), which was great for whales, although some email research in London had already told me that the weather would be turning about now, and that the whale-watching season was almost over.

This was a stark reminder that my first few weeks in North America were at the end of the summer, in sharp contrast to the remainder of the Trip, where hopefully perpetual summer would be the order of the day.

I also remembered reading years before that Vancouver was one of the world's most cosmopolitan cities, what with its situation on the western seaboard and a large immigrant Asian population from the other side of the Pacific. It sounded like it might be fun.

I felt I would have to be much less *English* when it came to making conversation with strangers, if I was to survive six months on my own, and meet some different and compelling people. My natural inbred reserve would have to be adapted to suit my ever-changing surroundings. This would have to apply especially to planes, as I was scheduled to be on forty-three of the things over the coming months, spending the equivalent of over a week in the air. It would get quite lonely if I let it.

With the Atlantic sparkling below me as the jumbo jet sped towards my first destination, I looked around me. Not an auspicious start. An empty seat next to me, an elderly chap who, it turned out, had been visiting his sister in London on the outside seat. Where was the business tycoon who could impress me with his tales of self-made millions? In Club Class, probably.

What about the sexy young traveller looking for someone to explore Vancouver with, and who thought my accent was just *so* cute? On a Greyhound bus right now, Russell.

So, I practised my small-talk skills on Ben, from Vancouver, and discovered that as long as you weren't talking to a Brit in the first place, people were genuinely open and unfailingly courteous. Ben was a mine of information about his home town – did I know, for example, that in certain months it was possible to ski in the mountains in the morning and then swim in the harbour in the afternoon?

As we flew in across the Canadian Rockies, dusk had already fallen, and the lights high up in the darkness gave just a hint that there were some *very* big mountains out there.

Once down into the harsh glare of a sparkling new airport, it became clear that Vancouver was typical of many of the airports I was to visit. Even though it was around 7 p.m. it seemed like a rather plush mall that was about to close. At least Canadian immigration was, I noticed with some relief, not the same tortuous exercise that I usually endured across the border in the USA. And my bags where all there, present and correct. One flight down, forty-two to go, I thought.

The city at first sight seemed to be much smaller than I had expected, and a short cab ride saw me dropped off at my impressive edifice of a hotel only an hour after I had landed.

The first hotel of the Trip was of gargantuan proportions, a huge monolith in the centre of downtown with what seemed umpteen bars and restaurants, although this being about 8 o'clock at night in North America most of them were already emptying – why is it that here people eat so much, and so early? Do they have to start at dawn just to fit it all in?

It also was a typical hotel from this part of the world, with an atrium full of facilities for it's guests, wonderfully attentive staff but tiny rooms that looked like they had been decorated by the visually impaired. Plenty of phones and Internet sockets, but a bathroom the size of a matchbox.

I decided that it was madness to unpack, as I intended to stay only a couple of nights in the city before moving on. A quick shower to dispel the jet lag, and off out exploring.

It turned out that I was staying very centrally, just off the main drag called Robson Street. This seemed to be wholly unexceptional and stereotypically North American – lots of familiar chain stores, some rather swish boutiques and what seemed like hundreds of places to eat. The only noticeable difference was that every other person appeared to be Japanese, or at least oriental of some kind.

After walking up and down Robson in search of something that was discernibly Canadian, I meandered a couple of blocks to Glenville Street, which was billed as the entertainment centre of the city.

For the first few hundred metres (and it is metres in Canada too, a sop, I imagine, to its francophone lobby along with dual labelling of place names and the like) it was full of bars and nightclubs, with a busy throng of drinkers and clubbers ready to noisily party their Tuesday night away.

Perhaps that's why they eat so early, I thought, because its just fuel to keep them going through the rest of the evening. I guess you tend not to dwell over Canadian cuisine. Even in London, a Tuesday night out would not be this raucous, this early.

I was also struck by how stark the contrasts are in North American cities. In the space of a few streets, Glenville changed from a party atmosphere, full of fashion-conscious, well-heeled Japanese yuppies, to a rundown ghetto of pawnshops, porn stores and greasy spoons, populated by some seriously mean looking hobos.

I finally succumbed to exhaustion a couple of hours later, reminding myself as I dozed off that it was important not to jump to conclusions based on one night and two streets. Tomorrow was another day. And one that would involve my first birds, I thought. I fell asleep, for the first time, but not the last, with an expectant grin on my face.

I had long empathised with Supertramp's 'Breakfast in America' track; there are few events in any visit to that continent that so neatly sum up its culture. Breakfast in Vancouver was no exception; a meal to be taken seriously, if only because of the vast array of choice available – half a dozen types of bread, muffins galore, lots of pork grilled or fried in different grisly ways, and all of it covered, of course, in wonderful rich, dark, pungent maple syrup, which often defied all attempts to wash it down with what they laughingly call 'coffee' in this part of the world. French influence has yet to permeate this far, I thought.

Massive on choice, big in quantity, low on quality and, as I discovered quickly, this being remote British Columbia, expensive to boot. The archetypal North American breakfast. I had arrived!

I figured the obvious place to begin tackling the all-important Bird List was the harbour. The bland streets leading down to the waterfront certainly didn't prepare me for the stunning view across to the mountains – one of two genuinely impressive things about Vancouver, the other being the gorgeous eponymous scarlet hues from the red maples in their fall foliage, which lit up the chill October day.

The city itself might not have a lot to commend it at first sight, but the Rockies framed against the sky, with the gold and russet forests separating them from the water, was a sight to lift the spirits on a drab fall day. For the first time, I felt that was far from home.

A Double-Crested Cormorant had the honour of being the first bird on the List, although you would never have known this as it sat nonchalantly around with its fellow cormorants on a harbour buoy. As with all new birding places, the adrenaline rose as I was able to pick off a dozen species quickly from the pier, the pick of which was a Belted Kingfisher, fishing from a post alongside.

And so, excited that finally the List was under way with sundry Loons, Ducks and Gulls, I walked through Stanley Park, a golden wonder of a place at this time of year, which dominated the junction between the city proper and the harbourside.

The park conveniently had a walkway around it, giving some great views of the cityscape looking back across the harbour. It also paid homage to the area's Native American forefathers, a heritage seemingly taken very seriously in this part of the world; the park was dotted with totems and carved figureheads, and these motifs were to be found all over the city.

An hour admiring the views had seen the day warm up considerably, and the populace of Vancouver were out enjoying the good weather whilst they could. It had yielded few new birds however, and I decided to dive into the woodlands in the hope of finding more.

The dripping trees and crisp air could have been anywhere in England, except for the unfamiliar bird calls deep within. They were, however, keeping themselves well hidden. After a while, I stood motionless in a sunlit glade, hoping that the birds would come to me.

And so, eventually, they did. After about ten minutes, the squeaking and chirruping seemed to be getting much closer, and then suddenly there I was, surrounded by a mixed feeding flock of Chickadees, Bushtits and both Golden and Ruby Crowned Kinglets (how fitting I thought, that they should mirror the colours of their forest). I felt elated and free from all the hassles in my life at home, which suddenly seemed so unimportant.

I would go on to see many more impressive bird sites, but few of them were as magical as standing there on my first morning away from home, with a blizzard of tiny unfamiliar birds all around me, for all the world as if I did not exist. I said to myself 'Well, if every day is going to provide birds as good as these, then the next six months are going to be even better than I thought'.

I spent a by now sunny afternoon in the Stanley Park Aquarium, an opportunity to see close up indigenous species like sea otters and beluga whales. And, with classic North American organisation and efficiency, it turned out to be a pleasant mixture of science and entertainment. The Canadians cannot make coffee, but they sure know how to showcase their wildlife.

I finished the day with wildlife of a different kind, but not before the first of many dozens of visits to an Internet café, until then a foreign concept to me. To some extent it still was a foreign concept, as I was the only non-Japanese in the place, and seemingly the only one not playing violent computer games. And there were two sets of keys, one in English, the others in Japanese.

Here began what turned out to be a marathon of email correspondence between Naaz and I. It was clear from her long note to me how upset she was by our strange parting in London, and it was hard to explain in writing just how wretched I felt about the whole thing too. But it did not stop us emoting like crazy in cyberspace at every opportunity over the next six months.

Curiously, Vancouver was to have the cheapest Internet rates of any country I visited, less than 50p an hour. There seemed to be no comparison in rates worldwide, ranging from these rates in Canada, or similar in somewhere like Ecuador and Vietnam, to more like £5 an hour in Key West in Florida. Ask me anything you like, I have become a world expert on Internet cafés around the globe.

I decided to end my second and last night in Vancouver with a visit to a lap-dancing club – after all, hadn't the Canadians invented the genre? Wasn't this therefore me sampling the local culture? That was my excuse anyway and I stuck to it.

Funnily enough, despite some remarkably sexy (and supple) women, the preponderance of silicon and the reek of commerciality acted like bromide in my tea and made the whole thing much less exciting than I was expecting . . . honest.

The next day I was up quickly, packed, breakfasted and on the road early. The plan was to drive through the rush-hour traffic to Horseshoe Bay, where the ferry leaves for 'the Island' as it's locally known, about 10 miles away.

I boarded the ferry as it sailed out into Vancouver Sound, the coasts of both the mainland and the island shrouded in autumnal mists. Halfway across, these had disappeared under the onslaught of a lively westerly wind, and the view, once again, was worth the price of entry alone. It was soon to be garnished by large flocks of Surf Scoter, a handsome black and white sea duck, which whizzed across the bow as we made our way during the crossing. It only looks a fraction of an inch on the map, and is testimony to another golden rule of travelling: distances abroad seem much bigger than at home.

This was also true of the Island, as the drive across the middle of it to the whale-watching centre of Tofino, on the Pacific coast, took up most of the rest of the day. Dull it was not, however, what with contending with the huge, loggers' trucks in the rain (this is one of the wettest places in Canada at the best of times), or marvelling at the fantastic scenery.

In fact, the mixture of lashing rain, huge crystal lakes and rolling mountains reminded me very much of the Scottish Highlands. That is, of course, until I saw the Bald Eagle.

I should tell you at this point that I had not been able to pin down this quintessentially American bird (in fact more numerous in Canada than the USA) on several previous trips, and that this was my number one target species in Canada.

As so often happens with birding, a surprise crops up when

you least expect it. I was rounding a bend, with a drowned forest on one side of the road and a mountain on the other, my head full of thoughts of what Naaz and JJ were getting up to that weekend, when suddenly an immature Bald Eagle left its perch on a dead tree and flew over the car, some 20 feet above my head. Now, *that's* the way to see a Bald Eagle!

As is also often the case, I could not then stop seeing them. I passed an adult bird sitting on an old eyrie shortly after, and a few days later on the ferry back to the mainland, I saw three birds in the space of five minutes. A fantastic spectacle, the huge black and white raptor framed against the blue mountain sky, or diving like a gull to pick up fish scraps from the water's surface.

I arrived in Tofino in an absolute deluge, and found a huge condo at an end-of-season knock-down rate. A quick investigation proved that there had been no whale-watching trips for several days, but that there was the prospect of a slight improvement in the weather the following morning before another front came in. I booked my ticket and hoped for the best.

Tofino turned out to be a well-heeled tourist town, a remote and soggy place full of wood cabins and prosperous looking resort hotels, just beginning to shut down for the winter. It was probably glorious in the summer, with canoeing, kayaking, camping, fishing, hunting, barbecuing and all. But here at what seemed like the end of the world, at the end of the season, I felt as though I was intruding as the only stranger for miles. The large numbers of Native Americans who seemed to populate the poorer end of town exacerbated the feeling that I was on the set of *Twin Peaks*. They looked somehow incongruous amidst the trappings of Western civilisation in their colourful shawls and hats.

And sure enough, when the weather was a little more forgiving the next morning, I was joined at the dockside by a grand total of two other tourists, Christian and Leila from, yes you've guessed it, Switzerland. The fact that they turned out to be really nice people was even more annoying. If only all the Swiss were a thoroughly nasty bunch!

The next twenty minutes were spent clambering into a variety of wet-weather clothing, including gloves, hats, boots – what do they know about the weather that I don't, I thought.

After our safety briefing, the little red zodiac sped out of Tofino's tiny harbour, with its idyllic backdrop of mountains and forest. I was grateful that yesterday's gale seemed to have dropped.

How wrong I was. Once we were out in the ocean, the waves were already a metre high, and looking out to the horizon you could see the ominous black cloud that was the approaching storm lurking like a malevolent beast.

I tried to ignore this and concentrate on the wildlife, as the boat scattered unfamiliar Pacific seabirds like Red Phalaropes and Pigeon Guillemots in all directions. We scudded across the molten ocean, and took refuge in a sheltered bay fringed by a crescent of sand and backed by thick conifer forest.

Suddenly the captain sighted a grey whale, one of a pod that had lingered here, feeding on shellfish on the sandy bottom of the shallow bays on their journey between the feeding grounds in the Arctic and the mating and breeding grounds in Mexico. Ten years before, I had been privileged to see these ugly yet fascinating creatures at the other end of their journey, and they had presented their newborn calves to us in a boat very similar to this one. Whale-watching was never less than a very humbling experience, and one you never forgot.

A grey whale looks nothing like the whales you expect to see – less a Moby Dick, more a long wet grey inner tube of an animal, encrusted around her face with white and orange barnacles, and with a foul-smelling spout like Billingsgate fish market if you got close enough. Not high in the sartorial stakes, but one of the most charismatic animals on earth. Despite her looking like a gargoyle of the sea, it was a privilege to share the ocean with her.

The fun really started when we left the whales, and ventured out into the bay to see a colony of Steller's sea lions, usually an Arctic species, on a nearby headland.

By now the wind was back close to full force, and the waves were rapidly increasing in height. Stationing myself in the front of the boat, for maximum visibility and birding opportunity, was becoming very much a mixed blessing. As we picked up speed, the zodiac, with only four people aboard, was skimming the crest of each wave comfortably enough, but then crashing down into the trough behind it with a crunch that made the fillings rattle in my teeth.

I tried bracing my feet firmly against the sides of the boat, and forcing my backside deep in to the seat, but it scarcely helped and the succession of wave impacts soon began to give me what felt like some kind of concussion. Before long I was *definitely* not enjoying the experience, face and glasses covered in salt, having to hang on to my hood to stop it viciously whipping my face, binoculars a lost cause and a throbbing headache competing with a churning stomach to see which could make me throw up first. I began to wish I were beneath the waves like the whales.

I discovered that the impact of each collision was lessened if I let out a grunt, much the same as when Monica Seles serves a tennis ball. Within minutes I was screaming with each wave, and if it were not for the fact that we were miles away from the safety of the harbour I would have voted to make a break for home. Plus, holding on to the rail for dear life, and with a roaring wind in our ears, the driver would never have heard me anyway.

We stopped by the sea lion colony, and quickly took some pictures in the gathering gloom, none of which were anything less than completely blurred in a swell that was now close to 10 feet. You could barely sit down, much less stand up, to take photos. The fact that I was able to tick a local specialty, Brandt's Cormorant, was scant consolation for the rising terror I was beginning to feel.

It was with some relief that eventually we put our backs to the wind, and headed for the shelter of a nearby bay, and the sanctuary of the harbour once again. A solitary Bald Eagle atop its eyrie at the mouth of the sound congratulated us on our return.

Once we were out of wet clothes and drinking hot chocolate back on dry land, it was exhilarating to ponder that we had been out on the last whale-watching expedition of the year, and in conditions that, if they had been prevalent at the start, would have seen us aborting the trip altogether. Now, *this* was the kind of adventure I had given everything up for, even if I was terrified throughout our hair-raising expedition.

So, mission one accomplished, I had one whale-watching trip under my belt and had seen some good birds to go with them. And, I had a jarred back to prove I'd had to work hard for them both.

However, the bird total was only thirty-nine, and I knew I had to keep moving and find new habitats if I was going to increase that. Plus, the local weather was going to be miserable for some time to come and, so, moving on was a priority.

A little research had told me that the other whale-watching centre on the island was the capital, Victoria, down in the south of the long, thin island that is Vancouver. As the weather was going to stop me doing much in the way of birding anyway, I decided to spend the whole of the next day driving south and arrive in Victoria in the late afternoon.

By the following morning, the storm was in full swing and the drive back through those mountains was eventful only for the sheer volume of water cascading down their sides and into their rivers.

It also never ceased to amaze me how, in the West at least, what looks like a remote settlement miles from anywhere on the map, can in fact be a modern town of some size, such is the march of civilisation. The other thing that occurred to me on that long drive was how much *space* Canada had – huge, cavernous, gloriously empty tracts where humans still ruled, but with a necessarily lighter touch. How different to the crowded little country I called home.

By the time I arrived in Victoria, the storm had passed and a splendid fall day was coming to an end. And 'splendid' was probably the word they would have used for it in Victoria; the place seemed quintessentially British, complete with its grand Victorian government buildings – a piece of the old British Empire still intact. It also seemed half dead – even on a Saturday afternoon the city, despite the suburban sprawl on the drive in, appeared to be only half full, as if the other half had got bored and gone off to the proper city back on the mainland.

After my first evening in Victoria I could understand why they might. Unlike Vancouver, there seemed to virtually no places of amusement, no bars and few restaurants, and half of those were already shut for the season.

Victoria did have one saving grace however. The Museum of British Colombia was a short walk from my hotel, and I could soon see why it was feted as one of the finest in Canada.

Half of it was devoted to tableaux depicting local wildlife, which it did with great aplomb (although none of the birds were tickable, unfortunately), and there were also some interesting scenes of the area with its prehistoric flora and fauna.

But the real gem was the section devoted to the local native Indian population; a plethora of totem poles, canoes, skins, weapons, implements, photographs and other artefacts. Although I thought I knew about how the 'red' Indian lived, this really opened my eyes and at the same time accentuated the perspective of time – whilst the white man had controlled this area for only 150 years, the red man had done so for thousands of years before him. What must the wildlife have been like when it truly *was* wild around here?

All of which made me yearn for another wild adventure. I did not have to wait long. The next whale-watching trip – aboard the (wait for it) *Prince of Whales* – was that very afternoon.

Here was an altogether different experience to the one in Tofino. The boat was big, it was full of Sunday afternoon adventurers and the weather was as clement as it had been ferocious further north.

This area was particularly good for orca (killer whale), a species I had never seen in the wild and was desperate to. However, the resident pod had not been seen for a while and the skipper said he felt they might have relocated further up the coast.

I was also hopeful of more bird species, once we had got further out to sea. I already realised how difficult that 1000 bird total was going to be; for instance, this was already my last opportunity to get birds in the northern Pacific, and of course I had only just scratched the surface.

Speeding along in the balmy afternoon sunshine, we located more seals and sea lions (this time it was calm enough for photos), but were a whale-free zone until another whale-watching boat radioed in a sighting some miles away.

We sped at full throttle across the flat ocean, and then idled at the spot it had been last seen. Suddenly a sharp black dorsal fin broke the surface behind us, followed by the sleek killing machine that is an orca. The boat exploded in a whirr of cameras and gasps of excitement as we all strived to get a view before he dived again.

This was a transient lone male animal, which had probably just cruised into the area in search of seal meat. In characteristic fashion, it breathed four or five times before diving, but we were able to follow the 'footprint' left by its tail flukes under the water, and an experienced skipper such as ours could pinpoint roughly where he would appear next.

If we lost him, as we did on a couple of occasions when the whale moved so swiftly that he left us behind, then he would eventually be picked up by another boat and we would get new directions and catch up. It occurred to me that if the ancient whalers had had this technology, then there would be no whales at all to watch today. A sobering thought.

We spend a fantastic thirty minutes tracking our orca, before moving on to pastures new, a very tame grey whale which had taken up residence in a bay close inshore – a bizarre sight, such a wild creature flanked by expensive bayside homes. She was, however, very happy with her lot and obligingly came so close to the boat that we had the dubious pleasure of smelling her oily 'breath' when she spouted.

There was still one remaining highlight for the ornithologists on the boat. I had already added another three species of Gull to the List, plus a Long Tailed Duck, when a pair of Harlequin Duck suddenly flew by. This was a real find. These birds spend the summer in the Arctic, and could not have been long in these more southerly waters. The male in particular is a stunning bird, with an elaborate blue and white head pattern that is unmistakable. This was, undoubtedly, the best bird of the Trip so far.

Having been on two whale trips, my mission to visit the island was fulfilled – the birds and the scenery a massive bonus. The journey back to the city of Vancouver gave me time to ponder on my photography strategy.

I had decided not to invest in an expensive camera. This was for the simple reason of portability and the fact that I wanted to spend more time with my binoculars than with any camera. I was acutely aware though that on these trips, if you don't have a photograph, it's almost as if you've never been in the first place. Crazy, I know, but true.

I expected to use up to fifty roles of film altogether and I was certainly not going to carry all *that* around with me – apart from being too heavy, it could get lost, damaged or stolen.

So I hit upon the idea of developing the films as I went, and sending them back to Naaz. Although this would use up holiday funds, it meant I got to see them soon after I'd taken them, I could edit out the ones I didn't want to keep there and then and I could also keep Naaz in touch with my adventures as I went along. Perfect.

I soon discovered that, even in the most remote places, you could find two things – a post office (where I could usually buy stamps to send home to James as well) and, much to my surprise, a film-developing facility.

I did a similar thing with bird field guides. I was determined to buy a field guide in each location, to maximise my chances of identifying species and, ultimately, add to my collection at home. (What next? A field guide about field guides?) The only trouble was, many of these were of necessity huge volumes, two or three times the size of this modest little work, for example. So, they too had to be mailed home, one by one, this time to my brother Tim. I wonder if the profits of the world's postal services have dipped since I got back.

I spent one last uneventful night in Vancouver, and then took an early morning cab to the palatial airport and a flight to Philadelphia in the USA, and the next stage of my adventure.

Canada had been a nice gentle introduction to travelling. It would get much more difficult before I got home, I told myself. As far as the Bird List was concerned, it was a question of 'never mind the width, feel the quality'.

I had only chalked up a total of fifty-one species in Canada, a function of the focus on whales, the weather and the time of year. This was just about on schedule (I needed to average a minimum of forty a week) but included few 'lifers' (birds I had never seen before, anywhere). I was already conscious that birding would be much more difficult in some other countries than it was in Canada and I needed to build up a big lead in those countries where transport and birding was easiest.

However, the memories of my first Bald Eagle, the pair of Harlequin Ducks, and the feeding flock in Stanley Park burns bright

to this day, and you could add the whales and the adven.
the storm off Tofino to that. If what I wanted from the Trip wa:
experiences, then I had certainly got off to one hell of a start.

BIRD COUNT: 51

Birding would be my priority in the USA. Although I had been to the USA some twenty times before, I had rarely done any concentrated birding. The focus of this part of the trip would be Cape May on the New Jersey coast, a noted migration hotspot. But the other thing I wanted to do in this part of the world was unlike any other activity on the whole trip.

I had long been fascinated by military history, originally a relic of my Dad's fascination for war and soldiers and battles and heroes. From an early age I had shared this. And way before I was daydreaming about girls or birds I was trying to imagine myself fighting for Cromwell in the Civil War, or perhaps in the trenches in Flanders. In the end, there wasn't a war I hadn't 'fought'.

This had never really gone away in my adult years and, in particular, I had become especially interested in the American Civil War. I remember collecting the bubble gum cards in the 1960s, and later there always seemed to be a drama on television re-enacting some episode from the war when I was a child.

And here I was in New Jersey, just a few hours' drive from the site of the war's most famous, bloody and, ultimately, pivotal battle, at Gettysburg in Pennsylvania. I knew that there was a museum there and that the battlefield was scarcely changed from the 1860s. It was too good an opportunity to miss, even if I had to give up some birding days to do it.

I also figured I would need some R & R before I left for South America, and thought that here was an excuse to spend some time in Florida – plenty of sun, a chance to see the art deco district in Miami South Beach and some birding in the Everglades for good measure. And, who knows, my first chance to find out if I could cut it in on my own, and meet some real live Americans -- maybe even the female of that species.

And so I flew into Philadelphia Airport, comforted by the fact that the country was so familiar to me, and excited that I would be able to fulfil a few ambitions all at once. I was also still giggling from the rather bizarre sight of my airline stewardess dressed as a witch, and the customs official in a Frankenstein costume – Americans take Halloween so *seriously*, don't they?

I had always liked the fact that, if you had a phone and a credit card, getting around in the USA was so easy – a great country, as long as you have money. So I wasn't fazed to find myself standing outside the airport in Philadelphia, in the dark, with nowhere to stay. A quick call and a courtesy bus saw me ensconced at a local motel, one of those soulless and cardboard American establishments where for a surprisingly few dollars you could have a close approximation of home comforts for a night, where I could phone Naaz easily, but where the coffee was assured to be even more repulsive than in Canada.

Alongside the cardboard motels were the requisite fast-food chains and diners selling cardboard food. I always make a point of eating in a diner at least once on any trip to the USA. It is a seminal experience, complete with grumpy waitresses, paunchy truck drivers, haggard sales executives, brash lighting and even more brightly coloured, but often tasteless, food. The menu is usually vast, and the food seems to be on the table before you've finished ordering it. But, the people are usually friendly – alarmingly for any Brit, they actually strike up polite conversation with you at the drop of a hat – and a three-course meal costs you less than a coffee and a croissant in Covent Garden in London.

My first surprise in the USA arrived the following morning, when back at the airport to rent a car, I managed to hire the last one remaining only by flirting and smiling sweetly at the black girl at the counter (in fact, in Philly, everybody seemed to be black.)

Duly equipped with my sporty little orange coupé I hit the road. Driving, like eating, in America is such a quintessentially . . . well . . . *American* experience. The roads are easy to find and easier to drive on than in the UK, there is more courtesy and less speeding, the billboards and malls which line the freeways look *exactly* like you expect them to and, with the election coverage on the radio as well, there was absolutely no question which country

I was in – unlike in Canada, where I could have been anywhere green and mountainous.

To add to my rising spirits, the fall weather in Northeastern USA was glorious, with bright sunshine and temperatures in the high 60s (°F). Despite getting lost (difficult I know in America) and spending an hour driving around across the state line in Delaware, the journey out to the coast and down to the Cape was an unexpected pleasure.

The countryside was positively groaning with provenance, vast wheatfields and orchards, little clapboard villages that reminded me of *The Waltons*, roadside stalls selling gigantic pumpkins, and election paraphernalia everywhere – this seemed to be a country where people actually cared who won, or perhaps it was just another excuse to erect a billboard on the roadside.

The second surprise of the day turned out to be the Cape itself. I had expected New Jersey to be dull and industrial, and the Cape to be like its equivalents in England – cold, windswept, with rubbish-strewn shingle and a rather dilapidated building masquerading as a bird observatory. How wrong could you be?

First, Cape May was actually a sizeable town and a rather prosperous one. Catering for the well-heeled romantic-weekend-away crowd, its genteel Main Street was lined with dozens of wonderful Victorian guest houses, all new paint and gabled roofs, and flying at least one ubiquitous Stars and Stripes. There were also hardly any cars, with the bicycle the preferred mode of transport.

Which was not to say that there were no people just because of the absence of cars – in this weather most of the guest houses were full, but I finally struck lucky at the Queen Victoria Hotel, a fantastic green extravaganza. The rooms had chintz wallpaper, big old ceiling fans, huge plump brass beds, and *no television*. It was like nowhere I'd stayed in the USA before, and all the better for it.

And then there was the Cape itself, a long curved spit of fine golden sand, with a picturesque lighthouse at one end. Near the town there was a reserve with two lagoons and some scrub. Alongside the lighthouse was an enormous car park, and a large viewing platform overlooking a lake on one side and the beach on the other. The observatory, as disappointing as the viewing platform was impressive, was a small nondescript hut which, in true American tradition, seemed chiefly concerned with selling Cape May merchandise (naturally, however, I did succumb).

Finally, some way beyond across the fields was another reserve of beachfront scrub called Highbee Beach. It was enormous area and I immediately regretted not having planned to spend longer here, just to see it all.

Cape May's key location on the eastern seaboard, just north of Delaware Bay, meant that birds migrating south from northern USA and Canada, on their way to the neo-tropic regions in Central and South America, would overfly the Cape and often pause to rest and feed before undertaking the journey over the expanse of the bay beyond.

This was probably the most famous of all US birding spots and, with its observatory nearby, was known as a 'must see' for every self-respecting British birder. It was most famous for its warbler migration at the end of September, when in the right weather conditions literally tens of thousands of small birds could turn up at the Cape – a 'fall' in birding parlance.

I had missed this, having been loved-up in London at the time, but I had now hit the purple patch as far as raptors (or birds of prey) were concerned. I had done some raptor migration work in Eilat years before and knew that it could be very exciting, as long as your identification skills were up to naming the 'dots in the sky' and your neck muscles and eyeballs could stand the stress of looking straight up for hours on end.

I spent the first afternoon at the lagoons near town, and could have been anywhere in, say, Norfolk, had the weather been worse and the birds different. Twenty-six new species fell immediately to my eager glare, including European wildfowl like Shoveler, Teal and Mute Swan which were hardly new to me but would be good ticks for the 1000 Bird List – my last chance to see species such as this before I departed south of the Equator. Undoubtedly, though, the highlight was a superb American Bittern flying low over the reeds, its big bow wings flapping soundlessly as it quartered the reed bed before disappearing within it.

However, I was really marking time until the morning, when I could get up on the viewing platform and count those raptors. At first light the next day I had breakfasted in America once again and was at the lighthouse for my first hawkwatch.

The weather had been great for the hawkwatch for a couple of days – a long depression followed by a clear, warm, calm period, where a bottleneck of birds could build up and then steam south on the high pressure which followed. The dynamics of bird migration never ceased to amaze me.

It usually took a couple of hours for the thermals to warm enough to attract the raptors, who would gain height on these warm breezes before suddenly pointing south and pretty much glide over the Cape, and across the bay to the Delaware shore on the other side. They would repeat this at suitable watch points all the way down the eastern seaboard of the USA, before one giant leap across the Gulf of Mexico and into Central and South America, where I hoped to see some of them again in Ecuador.

I therefore had time to scout the woodland around the lake, picking up Downy and Red Bellied Woodpeckers and a Northern Flicker, as well as clouds of Red Winged Blackbirds foraging in the reed beds, and being careful not to tread on Palm Warblers, which were so tame and seemingly under your feet at every turn.

By 10.30, I was back on station at the platform, and whilst waiting for the first birds was able to ponder on the differences between American birders and our own species back in the UK.

First, they looked different – usually bigger, older, more suntanned, grizzled even. There were not as many as I expected, although I guess the warbler migration was a bigger draw in September. They were unfailingly kitted out in various types of outdoor gear, as if a hurricane was expected at any moment, and had between them as fine a collection of expensive optics as you could see anywhere. All a far cry from the moth eaten and bedraggled bunch of Brits you might find at Spurn or Dungeness, our own rather more homespun versions of Cape May.

Second, there was none of the lighthearted banter that back home made even the most birdless twitch worthwhile; this lot were a serious bunch, driven it seemed only by the desire to let nothing pass unidentified in a bid to beat last year's total. As such, there was little of the bonhomie and information sharing that goes on at British reserves. This was free-market birding, with every person for him or herself. If you didn't spot something, they sure as hell weren't going to help you or show you a bird you may have missed.

This attitude was embodied by the official record keeper, an intense giant of a man who never used two words when one would do. He was responsible for the official count, and for updating the board in the car park, meticulous in its detail of the previous day's sightings. It seemed to be too much to ask him to be helpful or even civil with all these other responsibilities to take care of.

I should have known better, I suppose. In the spring I had been onto the Cape May birding website, appealing for advice. Several people had offered me help, including one who claimed to know all the local hotspots and who said he would take a day off to show me around.

All he wanted in return, he said, was for me to publicise his offer of expertise on the website, under his birding 'handle' of The UnaBirder (as in The UnaBomber, a maniacal domestic terrorist arrested a couple of years before).

Having done so quite innocently, I was then deluged by emails from other local birders, all warning me not to associate with this low life who had obviously done something to upset local birding protocol at some point. In the USA, even birders can be political, it seemed.

There were exceptions of course. As the first birds appeared as dots over the trees on the horizon, a huge jolly man by the name of Bill sat next to me and, being American, immediately acted as if I'd known him for years.

Bill had retired very early to Cape May to chill out and teach a little music. He didn't know much about birds, but came down to the hawkwatch every day to meet people and chat, and was as animated and cheerful as the true birders were serious and taciturn.

He was amazed that I had come this far to see some of 'his' birds. I felt like saying to him 'This is nothing mate, you should try driving from London to Yorkshire just to see one of your Green Herons'.

He laughed uproariously when I told him that I had given up my job primarily to go and see *lots* of other people's birds. He was one of many people I met along the way who impressed me with their kindness and who were genuinely envious of the risks I'd taken and the trip I'd planned.

Bill started to point up with his hammy fist, and try to identify the raptors that were now coming lower as they approached the Cape proper, and with it our viewing platform. As he squinted up through the glasses seemingly embedded in his flabby face, I began to realise that I knew more about the passing birds of prey than he did.

Over the next two hours, over 200 raptors flew over. Around half of these were Turkey Vultures, probably the most common of the US birds of prey. Most of the rest were either Sharp Shinned (or 'Sharpies') or the larger Red Tailed Hawks, and I quickly learnt to distinguish the Sparrowhawk-like fluttering flight of the Sharpies from the Buzzard – like soaring flight of the Red Tails. That was the easy part. Telling a Sharpie at distance from its larger cousin, the Northern Goshawk, was tougher, but not as tough as telling a Sharpie apart from its near relative, the Cooper's Hawk, (or 'Coop') at any distance.

Fortunately, the best birds of the day were easier to identify. A Northern Harrier glided menacingly along the beach, whilst a juvenile Golden Eagle drew appreciative gasps from the assembled grizzled and sunburnt throng.

However, for me the best bird was a bird I could have seen at home, but which had a very special significance – bird number 100 on the List – a Peregrine Falcon flying low and right overhead.

After a cholesterol-packed lunch, the afternoon was somewhat wasted getting lost amidst the tangles at Highbee Beach, with only a few birds for company. I had been warned not to stay out in this area after dark, and that it was a noted cruising location for homosexuals at night. An alarming couple of hours going around in circles as the sun set did nothing for my nerves, nor did finding that my car had been locked in the car park by an overzealous reserve warden. The area was beginning to fill up with moustachioed individuals in tight leather trousers by the time I found the warden again, and he mercifully set me and the car free.

I also spent the next morning at the hawkwatch, where birds

were thinner on the ground as the bird bottleneck eased. A couple of American Kestrels were welcome additions, as was the advice from Bill to go to Stone Point, a beachside town on the barrier islands some 10 miles or so to the north. But not before the bird of the day had been seen at very close range – an exhausted Virginia Rail, picked up by a British birder right in front of the platform, and photographically very obliging. Almost as impressive was a monarch butterfly, almost the size of a saucer, doggedly following the same migration route south. It's amazing to think that something so fragile can migrate from Canada to Mexico, where hundreds of thousands all congregate in the same patch of forest.

A variety of habitats are essential to any successful birding list, and so it proved at Stone Point, a relatively new development encroaching on some rather nice dunes and coastal mudflats – such is progress.

This gave me the chance to tick eight wader species, including the intricately titled Semi-Palmated Sandpiper (or 'Semi-P' – you can always rely on a birder for a great abbreviation) plus a rather nice Osprey fishing in the harbour itself. The flock of Black Skimmers on the beach outside my hotel was an added bonus.

The observatory may have been small, but at least it gave me the opportunity to buy into a couple of birding events. The first of these was the first owl watch of the season. For some twenty years, this husband and wife team had been trapping Owls on their routes south, and compiling invaluable data about their migration patterns. The more I find out about birding fieldwork, the more I realise how great a debt we owe to a small but fanatically dedicated team of amateurs around the globe, and the USA is no different.

A small group of chilly birders huddled together in the reserve car park, listening to the tape lure of a Saw-Whet Owl mournfully echoing around the dunes in the darkness. Eventually two rather bedraggled and sheepish owl watchers appeared from the dunes to explain that the night was too clear to bring in the Owls, which in such good conditions were more likely to pass straight through.

Just my luck, I thought, but it was once again proof of one of the unwritten rules of birding – most of the best birds are seen in awful weather. We must be so anti-social, praying for wind and rain whilst the rest of the world is hoping for sun and a calm breeze – bring on a storm to keep the birders happy!

That night we had to be content with hearing the two naturalists describe the migration cycle of the relevant Owl species, how they only migrated south from Canada only as far as the northern part of the USA, and how that was dependent on local weather and food conditions. These guys spent a large chunk of their marriage living out of a car, netting Owls in the dead of night in the middle of nowhere, if they were lucky – on most nights, like this one, there were simply no Owls at all.

The second bird trip, a walk in the local farmland the following morning, was more successful. The clear night had caused the first frost of the fall, and I had cause to be very grateful for my investment in a Cape May sweatshirt as we gathered in a woodland car park, stamping our feet and blowing clouds of hot air. So, this is how they all get so grizzled, I thought.

The clear weather conditions had seen a lot of birds arrive overnight, and immediately it was obvious that flocks of American Robins had been active before dawn; they were everywhere we looked on the frozen farmland, as were the flocks of Eastern Bluebirds overhead – unfortunately too high and too fast to pick out their exotic plumage on most occasions.

As the day began to warm, the woodland began to gently steam and drip. Birds began to emerge to forage on patches of sunlight – House Finch, four species of Sparrow, and Yellow Rumped Warblers yet to move south.

My favourites were a small party of cinnamon-coloured Cedar Waxwings, but it wouldn't have been birding without the one that got away, and the one lucky observer who glimpsed the Black Throated Blue Warbler made me very envious.

There are few birds so magical as the American warblers, tiny and often brightly coloured, which can regularly be 'pished' out of a bush for all to see. Unlike our own Warblers, most have a distinct autumn plumage, which presents a whole new set of identification hurdles. They have a special place in any birder's heart, and are one of the main reasons to go birding in the South American rainforest, where many of them winter. I made a mental note to go back to Cape May for the warbler migration one year.

It was with a heavy heart I left Cape May, still bathed in autumn sunshine. The bird count had risen to 122, a more than respectable start. I figured I had earned a night experiencing the sins of the flesh as well as the mind, and decided to break my journey to Gettysburg at the Las Vegas of the East Coast, Atlantic City.

This turned out to be the first mistake of the whole trip. I was prepared for Atlantic City to be different to the rural idyll I had just left, but I thought that it would be like Las Vegas – larger than life, brash and full of venal pleasures. I should have guessed I'd be wrong as I drove the ninety minutes north back up the coast. I began to have a sense of foreboding as the gigantic towers of the downtown hotels began to loom over the flat saltmarsh of the Jersey shore.

I needed to find a room for the night, and figured that even on a Saturday this would be no problem in the entertainment capital of the East Coast. Wrong. The very well appointed tourist office astride the freeway into Atlantic City politely rang around a few of the top hotels, and informed me that, this being the weekend, the city was full; all they had was a room at the Tropicana, for well over £100 a night.

'Blimey,' I thought 'this town must be *great* to get this popular', and promptly decided to break the budget, push the boat out and fully embrace the high-roller lifestyle. This of course was the first mistake; only the top hotels were actually full, and there were hundreds of tacky small motels that would have given me a bed for the night at a fraction of the cost. But this was by no means the main problem.

The *main* problem was that Atlantic City was officially the Armpit of the Western World. Brash it certainly was, but without any of the ritzy fantasy of its rival in the Nevada desert. This was a monument to bad taste. It seemed to be populated entirely by members of the audience from the *Jerry Springer Show*, all blue rinse and too much jewellery (and that was just the men), bulging guts and tattoos (and that included some of the women).

These poor unfortunates were bussed in from New York and Jersey by the thousands, to spend eight hours putting nickels in the fruit machines, before being bussed out again at 3 o'clock in the morning.

Recognising their marketing niche, the hotels squeezed gambling machines and blackjack parlours into every square inch of space. As a consequence, there were no great restaurants, only fast-food joints, and no lavish cabaret, just cheesy lounge muzak.

Sure, the hotels looked as if they were auditioning for a walk-on part in Las Vegas, each themed as a Wild West town or an Egyptian temple, but somehow in the chill breeze coming off the Atlantic, and knee-deep in white stilettos and baseball shirts, they never came close to making that particular grade.

It was one of the worst places of the entire Trip. As you can probably tell, I hated every minute of my eighteen hours in Atlantic City. I was quickly away from Armpit the next day, and took a leisurely drive through the ripening countryside back towards Philadelphia. I timed my journey to miss the gridlock that I suspected would arise on the road I needed, courtesy of the Eagle's match-up with my team, the New York Giants, and finally emerged on the western side of the city in the late afternoon, heading towards Gettysburg.

As in Canada, I was amazed by the enormity of distances in the USA. Philadelphia's suburbs seemed to go on for hours, and it was dusk before I hit the farm belt. And as I had been in New Jersey, I was surprised to find how rural a state Pennsylvania is. I had anticipated coal mines and factories, but what I got were patchwork fields, Dutch barns and charming little villages that looked as if they had been frozen in time.

The reason for this soon became clear. As I sauntered on through the twilight, I saw two dim lights ahead, belonging to a vehicle waiting to turn onto my road. I slowed down to give way, and was confronted by the extraordinary sight of a horse-drawn buggy, complete with candle lamplights and big black hood, and best of all driven by a character with a long beard, white shirt, braces and black hat. I was in Amish country, and the contrast with Armpit, less than a day's drive away, could not have been greater.

Deciding that budget food and accommodation would have to suffice after my expensive sojourn in Armpit, I stayed in a roadside motel and used the local diner, both of which reaffirmed my faith in the USA as great value outside the main towns. In particular, my first breakfast appointment with biscuits and gravy (or scones with packet sausage sauce as we would describe it) remains in my mind, not to say on my waistline. And, the price wouldn't have bought a packet of peanuts back in Armpit.

I got to the town of Gettysburg around mid-morning, found the museum and began my pilgrimage. It was partly the pageantry of the US Civil War – Blues versus Greys – that first caught my attention, but as I grew older I learnt that this war, one of the bloodiest in history, was fought at a key moment in military annals.

The tactics were still largely the same as had been used for a couple of hundred years in Europe, but the weaponry was busy evolving into the twentieth century, and this paradox was responsible for much of the bloodshed. The pathos and drama was increased further by the notion of the impoverished and rural South, fighting for a cause that was antiquated but that they still believed in, against the wealthy and industrial North, already well on the path to becoming the economic and military power that we know today.

Add to this the image of families fighting each other, of freed slaves escaping to the North and fighting against their old masters, and the fact that, against all odds, the Confederates in the South almost beat the Unionists in the North, and I was well and truly hooked.

The American Civil War was fought by soldiers only four generations before mine, and was the first war to receive significant media coverage. As a result, the uniforms, weapons, equipment and photographs were all in pristine condition, as befits a nation who cherish every bit of their short history and who know how to run a museum with more panache than many of their counterparts back in Europe.

At the time of the Civil War, the fashion was for painting Panoramas, or giant circular paintings, of big military battles (there is one at Waterloo I understand). The circular building next door housed the famous cylindrical painting encapsulating key moments of the three days of battle, albeit with a modicum of creative licence. It was a novel way to relive the agonies of the two armies locked in combat.

However, amidst the myriad of memorabilia, of replica banknotes, authentic Civil War barbecue sauce, copies of the Lincoln

Address and a mountain of books on the war, was a little gem that was to bring that historic battle to life for me. It was part map and part audiocassette, the idea being that you drove around the battlefield, stopping at key points and switching on the tape to follow the seesaw course of the seventy-two hours of the battle.

I spent the afternoon thus employed, revelling in the clear autumn sunshine, photographing the dozens of bold black statues commemorating heroic action by many of the regiments involved and imagining first hand (so to speak) how the battle that shaped the course of the war, and ultimately therefore the shape of the modern country, had unfolded.

Like all great battles – think of Waterloo, Balaclava, D-Day – this one was full of stories, of great pathos, tragedy and heroism, and I relived them all: the defence of Little Round Top, where Union soldiers of the First Maine fought off repeated attacks against the strategic hill by the rebels, finally resorting to hurling rocks when the ammunition ran out; how the Confederate General Robert E. Lee was let down on the first day by the ineptitude of his colonels, who failed to press home their early advantage, thereby surrendering the crucial high ground to an under-strength and badly organised enemy; and, of course, the infamous Pickett's Charge.

Both armies had spent two days fighting themselves to a standstill, and the logical course of action would usually have been to withdraw, conserve forces and live to fight another day. But Lee knew that he may never get his adversaries within range again, that this was the first (and last) major engagement of the war actually fought on Union soil, and that if they could emerge as victors, then Washington and ultimate glory were within their grasp. The war turned on the outcome at Gettysburg.

In desperation, Lee implored his last fresh troops, commanded by a southern gentlemen name of General George E. Pickett, to mount a near suicidal frontal assault on the centre of the Union line. And suicidal it most certainly was, I thought as I stood at the point in the line where the charge began, and looked a mile across the cornfields to where the Union line would have been. In the spirit of a bygone age, spurred on by now antiquated notions of honour and duty, the farm boys from Virginia and the Carolinas marched and died under the onslaught of cannon shells and rifle bullets from the opposing lines, slowing down to climb farm fences as they did so. They must have been very easy targets.

Tragically, some of the survivors did somehow manage to reach the Union lines, and even briefly pushed the defenders back before buckling under Union reinforcements. They retreated having suffered appalling losses (including just about every senior officer killed). It was an intensely moving spectacle, even for a foreigner nearly 140 years later. It had nothing to do with birds, but it was my favourite day of the whole of my stay in the USA.

The following morning I travelled back to Philadelphia with thousands of other early morning commuters, with thoughts of Florida sunshine on my mind. I gave back my orange coupé at the airport, intending to exchange it for a seat to Miami. But it was here, in the most technologically advanced nation on earth, that I suffered my only flight cock-up of the entire Trip.

A faulty plane meant that I had a six-hour wait for the next one south to Miami, but I managed to talk myself onto a plane going north to Boston, where a rapid connection would bring me south again and into Miami only an hour late.

I was amazed again at the way Americans used planes like we would a bus – those distances again – and in those days, a year before the attack on the Twin Towers, the plane was full.

As we flew north, I was lost in a reverie, thinking of the fantastically romantic weekend Naaz and I had shared only six weeks before in New York. How I missed her. And then, with an almost audible fanfare of trumpets, a gap in the clouds appeared momentarily outside my cabin window. Looking down, as if painted on the clouds, was a complete view of Manhattan Island, with its tiny Empire State Building and miniature World Trade Centre. In a few seconds it was gone, swallowed up again by the clouds, but it sent shivers up and down my spine. And, in that second, I knew I would see Naaz again, and sooner rather than later.

I also flew over Cape May on the way south, mirroring the migration flight of all the birds I had seen only days before. Seen from the air, that entire vast area looked so insignificant, and it struck me that it was a miracle that any birds were seen at all by those on the ground, as the point itself hardly looked a birding hotspot from several miles above it.

As the plane banked above Miami, the setting sun shone on the seemingly endless sheen of water that seemed to pockmark the entire southern half of the state. We circled the city right out over the Everglades, and it was possible to see not only all the lakes and marshes amongst the grasslands, but also how agriculture and suburbia was constantly nibbling away at the edges of this once great wilderness.

Miami airport was a complete zoo and, as such, an accurate reflection of the city it services. It was very humid, chaotic and crowded, and if I didn't know better I would have sworn that we were somewhere in Latin America – the predominant language was Spanish and most of the people in the airport seemed to be either Hispanic or black.

The biggest difference between south Florida and my first week in the USA was immediately apparent. Despite the great autumn weather further north, the temperature down here was some 25 °F higher, with nearly 100 per cent humidity. The political temperature was also high, with this being Election Day, and without any hint of the drama that was to be laid at Florida's door in the coming days.

I had spent little time in Miami on my previous visits to Florida with Sue and James, especially as at that time tourists were warned off from staying in the city for fear of mugging or car-jacking. Since then, Miami had cleaned up its act, and nowhere more so than in South Beach, a suburb out beyond the city proper and officially in the city of Miami Beach.

Famous as the art deco centre of the area, South Beach had formerly been notorious as much for its winos and drug addicts as its faded grandeur and artist colony. A major clean-up by the city had seen its rebirth, and I made this my first stop, excited by the thought of some tropical consumerism and a pastel paradise.

It was dark by the time I rented my car, and a series of frantic U-turns finally saw me on the freeway across the harbour, past the gigantic cruise ships docked there, and within sight of my first palm trees amidst the neon strip that was Miami South Beach.

Collins Avenue is at the centre of the art deco district, and contains all the fancy hotels in the area. Adjacent is Ocean Boulevard, right on the beach, with all the bars and restaurants to cater for the tourists staying on Collins. I had booked into the nearby Holiday Inn, which turned out to be some seventeen blocks down Collins Avenue, far enough at any rate to still be waiting for its own facelift. Nevertheless, I determined to act the tourist for a few days, as it would be my last chance for some considerable time.

I soon spotted the flaw in my plan. It was tough being a tourist by yourself, especially when everyone else seemed to be in groups of happy, fun-seeking partygoers. I pressed on, determined not to let loneliness ruin my time here, but this was easier said than done and it took me the length of time it takes to eat a steak before I was wishing that Naaz could have been there to share it with me.

A very supple black Trinidadian teenager writhing in my lap was probably not the best antidote to the homesick blues, but somewhat inevitably that is how I spent the rest of the evening, in a rip-off lap-dancing club on Collins. At least in there I had one thing in common with everybody else, which I did not have out on the street – in here, we were *all* alone. Trouble is, I felt even more lonely afterwards, not to mention lighter in the pocket.

I awoke the following day to the furore over an election too close to call, and the media circus had begun to roll into Miami already as the squabble over recounts was just beginning. Ironic, I thought, that an election where the turnout was almost the lowest in the democratic world, should suddenly spark so much interest.

I decided to spend the day chilling, as the first two weeks had seen me seldom spend more than one night in the same place. South Beach was a curious place, with some fabulously smart pink or green art deco hotels typically stuck next to some dilapidated dime store. The prices here were the highest of any place I stayed on the entire trip, as befits one of the most stylish resorts in the USA.

I cruised Ocean Boulevard, gazing at the huge open-top Cadillacs, the roller-skating beach babes and the iron-pumping gays on the beach. I had never seen so much silicon in one place before.

For all the bars and cafés I sampled over the next few days,

South Beach was an uncannily empty place to be a lone tourist, and I found myself on the phone to Naaz that night vowing to fly home – for the weekend no less – whilst I still was relatively close to the UK.

To her credit, Naaz persuaded me not to push the button on the option I had to fly to Heathrow the following day. She said that it would be a coward's way out, and I had to admit that it would do little for the integrity of my Trip.

However, I also knew that what she meant, but could not say, was that she felt she now had to concentrate on her relationship with JJ, and my return would inevitably distract her. I never told her that I actually tried to firm up that flight later that night, but that it had been sold in the mean time. Who knows what might have happened if I had come back after only two weeks. Would I have had the guts to return to my itinerary? Would I now be married to Naaz ? In the end, fate lent a hand and made my decision for me.

Instead of a flight to London therefore, the following day saw me driving down Highway One, en route to the Florida Keys. My destination was Key West, which I remembered from my first visit, some ten years previously with Sue, as a bohemian and funky town, unlike anywhere else in the USA. Perhaps this wasn't that surprising, as Key West is actually nearer to Cuba than to Miami, and revels in its self-styled independence, to the extent of having its own flag of the 'Conch Republic'.

I struggled to find accommodation having arrived on the same weekend as the World Powerboat Championships, and I ended up paying a fortune to stay in a vast apartment in the centre of town. Key West had changed only a little in the past ten years. I remembered it as a big gay centre, and this had seemed to have been toned down a little or perhaps just submerged beneath the hundreds of middle-aged partygoers down for the championships. I had never seen so many drunken Americans in all my life, and walking down Duvall Street to see the sunset at Mallory Square was to see all human life portrayed – drag queens, transvestites, bikers, rockers, hookers, beach folk and plain old-fashioned tourists mixed together in an exotic tropical cocktail.

I still struggled to enjoy it all, however. I lapped up the sunshine, ate and drank fantastically, and shopped for Christmas gifts with abandon, but by myself it just never added up. After a cocktail or two, I found a willing victim to chat up in a bar, but even this was curtailed when her boyfriend suddenly showed up.

So I did what I always did whenever I was single and lonely, and went birding. Here I felt more comfortable, as birding was by design a largely solitary pastime, and a field I felt comfortable in. I could at least enjoy this as much as the next guy, unlike the non-stop festival on Duvall Street.

Earlier that year, Anthony, a birding friend of mine, had spent an incredible day birding at Fort Jefferson, a remote outpost in the Dry Tortugas, about an hour by seaplane from Key West.

Anthony had been lucky enough to be at Jefferson in absolutely perfect conditions, with American warblers returning north from the neo-tropics, and flopping down virtually at his feet, exhausted and starving. He had marvelled at the grisly sight of an assembled throng of raptors, picking off the hordes of gaudy warblers one by one until the Fort was suddenly empty again, in a ritual repeated on islands all over the Caribbean during migration since time immemorial. He had first-class, point-blank views of dozens of new species, both the predators and the prey, and it had really whet my appetite.

I already knew that my visit was out of season, and that the Fort would be much quieter in November than in April, but it was a pilgrimage I was excited to undertake, especially as it meant my first flight on a seaplane and it got me away from the nauseating happiness back on Key West for the day. Also, if I was to reach the magic 1000 figure, it was about time I got back to finding birds.

I pretty much had the seaplane to myself, and was surprised to find that it actually took off on wheels from the airport, as opposed to the sea. The trip out across the Caribbean took us over the powerboat races, giving us fantastic panoramic views of the sleek machines denied to the drunken hoards on the shorefront. We also spied a couple of old shipwrecks, their masts still pointing skeletally above the shallow waters many years on.

A fifty-minute flight saw us circling Fort Jefferson, its red-brick hexagonal shape totally dominating the tiny island it was built on. Its most important function, apart from as a haven for migrating birds, had been as a defence against the Spanish in Cuba at the turn of the last century.

As we bumped over the waves and came to rest on the beach, it struck me that I would be going to many exotic places in the coming months, but perhaps nowhere quite as remote as this. In the USA of course, being remote was never going to stop the ubiquitous tourists, and the grounds outside the Fort already had a few campers, their yachts tied up in the tiny harbour.

Around this harbour where a series of posts, each providing a handy fishing post for its bird – Brown Pelicans or Sandwich and Royal Terns or Willets. A quick walk around the island revealed that the utter remoteness of the island meant that the bird life, like Ruddy Turnstones and Sanderlings, were as tame as anything, and a taste of what was to come in the Galapagos later that month.

The interior of the Fort was not festooned with birds as it had been on Anthony's visit, of course, but as it had the only drinking water on the island it did attract the few passerine migrants still journeying south – an Eastern Phoebe, Eastern Wood Pewee (US birds do have such convoluted names) and a Blue Grosbeak, whilst a Common Nighthawk was still hawking for insects along the line of the Fort's ramparts.

Jefferson had placed birding back on my radar with a vengeance, and the following morning saw me heading back up Highway One, across the Keys once again towards the Everglades and more American birds. I had expected this day to be dull, most of it spent in the car crossing what the Indians locally had termed 'The River of Grass', but I was in for a nice surprise.

After only a few miles back towards the mainland I passed a sign on the highway entitled 'Keys Hawk Watch'. Scarcely believing my luck, I pulled into the adjoining campground and joined a clutch of birders atop the old lighthouse.

I wasn't even aware that there *was* a version of the hawkwatch down here, but it certainly made sense, as the string of islands leading out across Florida Bay represented the very last landfall before the trek across the Gulf of Mexico for the type of birds I'd been watching hundreds of miles further north only a week before. In fact, I mused, some of them might even be the very same birds.

The Keys crowd was very different to its northern counterpart –

younger, more laid back, happy to help this foreign birder and fascinated by my birding odyssey. Plus the birds were much lower than at Cape May and, on this day at least, the last of the migration season, more plentiful.

I was there over two hours in the end, and identified a dozen species of raptor, including Swainson's and Short Tailed Hawks and a Merlin, all of which were not seen in New Jersey. Most noteworthy of all, though, were the two Bald Eagles, my first in the USA.

The local warden was very generous with his tips on where to find birds in the Everglades, and by the time the migration had begun to tail off in the early afternoon I was on my way again.

Spending the night in Florida City (sounds exotic doesn't it, but think of the M25 with more humidity and you'll be close), I then took the road north again intending to link up with 'Alligator Alley', the road that bisects the Everglades.

For anyone aware of the history of this once vast swamp, this drive was fairly depressing. Once the Everglades (a misnomer by the English if there ever was one – there is barely a tree in it) stretched from Lake Okeechobee in the north down to the sea, a huge, very slow-moving river delta, draining the whole of the southern part of the state into the Gulf of Mexico. Ignored by first the Spanish and then the Americans, it was left to the native Indians until it fell victim to the land rush in the early 1900s for agriculture, and in the 1930s for tourism.

Today, it is less than half its original size, and so much pressure has been put on the local water sources by the surrounding cities, resorts and farms that it barely flows anywhere any more, a problem made worse by the high concentrates of agro-chemicals that flow into it from the surrounding farmland. The swampland that is left is criss-crossed by dykes and pipelines around its edges, reducing the really wild places to a handful of reserves (or 'preserves' as the Americans would have it).

That is not to say the modern-day Everglades are small – it still takes over three hours to drive across it and its wooded sister, The Big Cypress Preserve – or that it is devoid of bird life. The Fish and Wildlife-service has designed some first-class reserves, which usually hold masses of birds despite being a little 'dude' in birding-speak

– the tourists are shipped around by a mini-train at Shark Valley for example (which is neither a valley nor has any sharks), whilst at another the visitors gawp at the huge alligators from carefully laid-out boardwalks.

Using a map picked up at a Ranger Station, I stopped off at all the reserves on the Tiamiami Trail, and duly logged Everglade specialities like Tri-coloured, Black Crowned Night and American Green Herons, Glossy Ibis, Roseate Spoonbill, Wood Stork and American Avocet. A journey further south to the coastal town of Flamingo also provided one of the few places where both Brown and American White Pelicans can be found together. Like many birding places in the USA, the facilities were first class and the birds incredibly tame, and so viewing the marsh birds at point-blank range was easy and well within the range of my modest little camera.

The best bird of the day, however, was the Snail Kite, a large black raptor with a distinctive white rump and a very particular appetite. This trademark bird of the Everglades could be seen from the road, or the boardwalks, quartering the grasslands, looking for the large snails which cling to the stems and carefully picking them off with its talons before winkling out the meat with its oddly delicate curved, pointed bill.

What a terrible shame that the local Park Rangers told me they estimated that bird life in the Everglades was only 10 per cent of the size it was a 100 years before – what a magnificent spectacle that must have been.

The trip to the Everglades also reminded me that there is always a day on any trip abroad when you first realise that you have come unprepared. That night, as dozens of mosquito bites on my legs began to slowly drive me insane, I realised what I had forgotten to buy at the pharmacy.

The following day saw me head across the Everglades, intending to spend the night at Naples on the southwest coast, an area I had never been to. I stopped at lunchtime for an Indian, in this case *Native American* Indian, a roadside restaurant serving speciality dishes as eaten by the locals.

This turned out to be a mountain of exotic foods, and all of it deep fried – frogs' legs, alligator, catfish and cornbread. At about a million calories a plate, it was easy to see why life expectancy amongst the Native Americans was so low, although many of them had long since made a fortune from building casinos on their reserves and were probably living in South Beach and feasting on lobsters and steaks. Either way, it was one of the most unusual meals of the Trip.

The List had grown by barely forty species since my arrival in Florida, and so the trip out to the coast was designed to pick up some shorebirds at a couple of spots that had been recommended to me and, hopefully, to see some manatee – the endangered sea mammal that comes to enjoy the warm Florida waters in the winter time.

There was little else to enjoy in this part of the state. The 40 miles or so between the twin cities of Naples and Fort Myers were one huge long stretch of malls, condos and car showrooms. This may have been the American Dream for many, but for me the thought of living in this turgidly commercial environment was more like a nightmare. The idea of spending *my* twilight years (average age seemed to be about sixty) driving very, very slowly from one shop to another via the nearest restaurant nearly made me throw up my deep-fried alligator.

The day after, the first overnight rain since Canada saw me searching fruitlessly for a manatee at Fort Myers – the temperature in the ocean was not yet cold enough to bring them inshore, apparently – before heading north, grateful to leave the city behind.

On a ranger's recommendation I ended up on Marco Island, one of a number of offshore islands, and tramped along the seashore to Tigertail Beach. (Have you noticed how they are never called something like 'Jaywick Sands' or 'Cromer' in the USA?)

Here at Tigertail were a number of sandbanks and creeks and, despite being near the centre of a tourist resort, the bird life was impressive. Reddish Egret (a local speciality), and standard shorebirds like Piping Plover, Lesser Yellowlegs, Long Billed Dowitcher and Western Sandpiper were all quickly mopped up, along with Florida Scrub Jay and a couple of inshore bottlenose dolphins at the delightful local preserve at Rookery Bay.

There was one more place I still needed to go to on this coast, Sanibel Island, a well-known salubrious resort further north, where not only would I be able to put down roots for a few days, but also visit the famous local reserve called Ding Darling after a local naturalist.

Once again I managed to rent a condominium that was several sizes too large for me, but at least it gave me a chance to relax in wonderful surroundings. Sanibel was a beautiful little island, full of funky beachfront houses and upmarket resort complexes.

For the first time I finally managed to chill a little, and spent some time soaking up a few rays of sun, mailing Christmas presents home (knowing I probably couldn't do that from South America, at least not easily) and reading. The local restaurants also took up some of my time, and I spent a memorable evening with a couple of political lobbyists from Boston, down to cover the ongoing recounts that looked to be gerrymandering Bush into power, and a very fine steak washed down with a bottle of Merlot.

At last I seemed to be feeling more comfortable with leaving my life behind me, after more than three weeks of more or less constant travelling. It helped that I could lose myself in the birds that were present everywhere I went, and I was in constant touch with Naaz, so even this was becoming more bearable.

Ironically, I made my first call home to my Mum and Dad, and a mother's intuition being what it is, she could sense that I had had a hard time in Florida by myself. I had to work really hard to persuade her that everything was OK, albeit after weeks of soul searching of my own, and despite the realisation that in hindsight a week less in Florida by myself would have been ideal.

The Ding Darling reserve turned out to be a great spot, few new birds but lots more of the ones I'd seen in the Everglades – dozens of Reddish Egrets, flocks of Roseate Spoonbills and Snowy Egrets that would almost let you touch them they were so tame. At migration time, the lagoons and marshes would have been laden with bird life, but in my more mellow frame of mind a casual drive around the reserve (What, walk in America? You must be joking!) each morning was still a wonderful way to start the day despite the lack of new birds.

With two days left until I left for Ecuador, I felt I should go back to South Beach to try and enjoy it more than I had the first time. I drove back across the Everglades, navigated my way through the labyrinthine freeway system in Miami and finally found myself in South Beach again, and this time booked into one of the swish, art deco, funky hotels at the right end of Collins Avenue, a purple and cream extravaganza called The Shelley Hotel.

This was more like it, and to hell with the expense. That night I even met a couple of real, live, friendly Americans. Fuelled by a couple of free happy-hour drinks at the dinky little hotel bar, I made a bee-line for a busty blond sitting by herself on the terrace outside.

Alice was all New York – thick Bronx accent, big boobs, big hair, big attitude, big drinker, smoked like a chimney and swore like a trooper. Unfortunately she was with Doug, a quietly spoken, married, long- time school friend of hers, who had treated her to a weekend in South Beach for her birthday without his wife knowing. Alice told me quietly that Doug's wife wouldn't have approved. As I looked down at her impressive cleavage whilst she whispered this to me, I could well understand why.

Alice and Doug kind of adopted me for the next couple of nights. I finally felt at home, drinking a different cocktail in each beachfront bar and smoking cigars and telling dirty jokes until 3 o'clock in the morning. Finally I felt as though *I* was one of the crowd, *I* was having a great time. I was no longer on the outside and at last I had seen South Beach as it was meant to be seen – through the bottom of a frozen Margarita.

After three weeks my US journey had finally come to an end. Despite my many trips to the country, much of what I had seen surprised me this time: how wonderful the Pennsylvania countryside could look in its autumn finery, the great birding at the hawkwatches, the classy resort at Sanibel, but also how it was not difficult to despise parts of Florida, the horror that was Atlantic City and how easy it was to be alone in a resort full of people. A fantastic country full of contrasts, that much hadn't changed at least.

The List had only added another 122 species in the three weeks I was in the USA, partly due to me attempting to relax, and also because some of the Canadian birds had overlapped with those in the USA. The overall total of 173 was still just about on schedule to reach the magical 1000 figure, although all of my efforts in the

USA had still only yielded a dozen 'lifers' – one less than the week I spent in Canada, or about the same as a good year in the UK.

I sat in the departure lounge in Miami airport, and pondered on my first four weeks away. Exciting, yes. Enjoyable, well mostly, but all very *comfortable*, this being civilized North America. I had little doubt that what lay ahead of me in Ecuador and beyond would be very, very different indeed.

BIRD COUNT: 173

4 The *Cachalote* : The Galapagos Islands, Ecuador, November

The flight to Quito had the air of flying into the unknown. South America was a complete mystery to me, a place I had never been, where little English was spoken and the people and customs unfamiliar. My adventure starts here, I thought.

As we flew in I could see the surrounding mountains on all sides, with the sprawling capital of Ecuador laid out along the valley. My neighbour, an ex-USAF colonel who had lived for years in Suffolk, told me that Quito was the world's second highest capital, after La Paz in Bolivia, and explained that he was working in Quito as a consultant to the security services. I should have guessed what that meant.

The interminable wait at Customs, where in South America the amount of officialdom seems to be in direct proportion to the amount of gold braid and badges on the uniforms, delayed me long enough to see my baggage pass through, always a welcome surprise given the chaos in an airport like Quito.

The contrast between your typical Hispanics in Ecuador, compared with Miami, was immediately obvious; rather than being of Cuban descent, the vast majority of the throng at the airport bore the high cheekbones and slanted eyes of the Andean tribes, similar to the Inca of the past.

As usual in the Third World, everybody was a taxi driver looking for my business, and it was some time before I was able to locate my guide for the duration, a charming, smiley Indian named Rodrigo.

On the ride into the city Rodrigo explained that I had two problems to be aware of. The first of these was pick-pocketing, a standard warning to all tourists everywhere, but made worse by the immediate financial circumstances. The government, in a bid to control rampant inflation, had just adopted the US dollar as the national currency instead of the peso. Although this had pegged price increases, it had thrown hundreds of thousands of people out of work, and the streets were now full of beggars and street vendors, whilst the Spanish Embassy was besieged every morning by hopeful souls seeking a visa to emigrate.

The second problem was one of a more enduring nature – altitude. At over 3000 metres Quito's atmosphere had less oxygen than most Westerners were used to, and the effects of altitude sickness could ruin your whole trip. Rodrigo advised sucking a sugar sweet if I felt unwell.

My first real shock in Ecuador was pulling up outside my hotel. It was a charming little town house, with a small courtyard, but was dominated by a massive steel gate, which obscured the entire frontage from the street. And what was worse, the gate was opened by a moustachioed paramilitary guard, complete with pump-action shotgun (every home should have one, at least in Quito) and ravenous snarling Alsatian. A fine welcome to what I had assumed was a peaceful country.

I had a day to kill in Quito – a poor choice of words given the amount of firearms I was to see, with a guard in every shop, bank and restaurant – before flying out to the Galapagos. With my heart in my mouth and a pocketful of sweets I was up as early as the jet lag would allow.

First impressions confirmed that Quito was very different to anywhere I had ever been before. The most obvious evidence of this was all around – huge, towering mountains on either side of the city, which you had to crane your neck to see the summit of – and the *valley* was at over 3000 metres, so God knows how high the actual mountains were. Look up from any street in Quito, and you are guaranteed to see a mountain, and a big one at that.

The mountains reinforced the notion that you were indeed high up, but such was the effect of altitude on that first day in Quito, that you really didn't need reminding. I can best describe this by saying that it was like going without sleep for forty-eight hours – a deep fug that penetrated the brain and left you with a nagging headache by evening, and a listlessness and general fatigue that meant walking any distance would need a rest soon after. Life seemed to amble on at a gentle pace therefore, and was seen through bleary eyes that seemed to want to close at every second. Perversely, one of the side effects of this was lurid, technicolour dreams of the most surreal nature, as if to destroy the rest your body perpetually seemed to crave.

Quito as a city seemed largely characterless, at least in the commercial and shopping districts. A few high rises were set next to streets of hand-to-mouth stores of typical cottage industries. The traffic was appalling (as it always seems to be in the Third World), with lumbering buses that were always full to overcrowding and farm trucks belching out clouds of acrid black smoke.

However, it did seem to have an unending supply of three commodities. The first was dust, which permeated every street and building. The second was travel companies, all seemingly selling identical tours to the Galapagos, the rainforest or, especially, trekking in the Andes. Thirdly, every other shop was an Internet café, where legions of backpackers from around the world congregated to email home for the ridiculously low sum of around a $1 an hour – or about a tenth of what it cost in somewhere like Key West.

In fact, whatever the nondescript nature of Quito, you could not fault its prices, inflation adjusted or otherwise. I ate lunch early on the first day, trying to rest my throbbing head and aching limbs, and struggled to pay more than about $5. It seemed so wrong that we Westerners could afford to eat handsomely for not much more than a pint of beer in the West End of London, whilst you were continually fighting off overtures from beggars who obviously had nothing. The more enterprising were camped on the street, selling everything from toothpaste to shoelaces to chewing gum, whilst others would try to get you to part with a few cents by sending over a ragamuffin child on a charm offensive.

Despite all the armed guards, I never saw any trouble on the streets, but it did look as though security, tourism and begging were the only growth industries in Quito.

My only task that first day was to find a field guide on Ecuadorian birds, since I had failed to do so in the USA. I was directed to Libri Mundi, the most notable bookshop in Quito, but its natural history section seemed confined to coffee-table picture books with colourful photographs of the reef in the Galapagos.

I had more luck in the tourist quarter, where I found a store specialising in wildlife trips, with a small library which had books for sale. There seemed to be no books devoted to the whole of Ecuador (although one has since been published), and the nearest equivalent was a huge volume of the *Birds of Columbia* – too heavy, not relevant enough and extortionately expensive at $25 a copy. You could eat for a week in Quito for that.

I made do with a book about the birds of the Galapagos, knowing that would take care of my immediate requirements, and trusting that I would be able to beg or borrow a field guide later in the trip for the other areas I was to visit.

For any wildlife enthusiast the Galapagos Islands have a special cache. Not only was this where Darwin's theories on evolution were formed, but it was one of the few places where a near pristine environment could still be found. In addition, its very remoteness meant that the animals would be very tame, and that many had evolved into species uniquely adapted to their environment and found nowhere else on earth, including Flightless Cormorants, marine iguanas and a species of Penguin which was easily the most northerly in the entire globe, thanks to the cold water Humboldt Current around the islands. Penguins on the Equator – who would have believed it!

I was looking forward to my week touring the islands with immense anticipation, not least because I counted this as actually *travelling* for the first time, and not the tourism I felt I had done up until that point.

I was up absurdly early the next morning to avoid the traffic congestion caused by the visit of the President to Quito, and had a breakfast time rendezvous with my companions for the next week: a group of Brits travelling around South America and finishing their tour in Ecuador; another eccentric Brit, living in Milan, starting his own world tour in the Galapagos; a solitary German and a funky Jewish ex-hippy couple from San Francisco.

The start of these little adventures was always a little strange; nine individuals with little in common except an itinerary, thrown together on a small boat for a week. However, I relished the thought of getting to know my new travel companions, telling myself that this was part of the adventure. Having lived my entire life in a cocoon where

I only ever drank, ate, worked and played with people in the same profession, it was a welcome change.

Our plane stopped in Guayaqull, Ecuador's large commercial port and industrial centre, before crossing the 1000 kilometres across the Pacific and landing at Baltra, a small windswept island in the north of the Galapagos archipelago.

Here we met Sonia, our larger-than-life guide, an Italian who had found herself in the Islands via a number of colourful love affairs, which had left her with several extra languages and a coveted permit as a tourist guide. If there was any doubt that we would take some time to bond, Sonia's loud voice and rumbustious good humour soon dispelled them. She was larger than life in every way, with a deep tan, extraordinarily voluptuous figure and a matriarchal attitude only a Sicilian can carry off.

As our coach descended from the airport to the harbour, I could feel the excitement rise in my chest and we all strained our eyes to be the first to spot the *Cachalote*, our vessel for the duration.

I had had an affinity with boats ever since my visit to see the whales in Mexico some ten years before. There was something about waking every morning and being surrounded by the open sea, about smelling the salt and feeling the sea breeze, about watching the birds and dolphins from the deck, and spending all day barefoot and in just a pair of shorts. Although I was less than comfortable being under the ocean, being *on* the surface of it mesmerised me, knowing that it was the least explored part of our planet and that anything might appear at any time.

Fond memories of the whaling trip had led to me deliberately booking onto a *proper* boat – that is to say, one with a sail and that looked like a boat rather than the floating gin-palace that seemed to be the norm – and I was not to be disappointed.

The *Cachalote* was built in 1971 and was reaching the end of her serviceable life as a touring ship. None of the crew wanted to lose her, despite the fact that she was older and a little cramped compared with most of her more modern rivals. I fell in love with her immediately – she had portholes not windows, brass fittings, two masts and a cosy salon. The salon was where everyone congregated since the *Cachalote*'s cabins were so tiny that to stay in them for anything other than sleeping was to risk instant claustrophobia. Navigating the ladders to get from your cabin to the salon whilst at sea became an everyday challenge.

There were five crew who spoke little English between them, and the most important member was not the Captain but the cook who, despite having a tiny galley at the rear of the wheelhouse, conjured up three immense meals daily. I thanked God that these tours were aimed at the US market and so the food was always in plentiful supply – as befits appetites sharpened by the sea air.

I also got very lucky as we were introduced around the boat – as one of the single guys aboard, it was *me* that got a cabin to myself. This was in the bow of the boat, and gave me more space despite the curve of the prow in the cabin – although *more* space was a relative term, since there was just enough floor space to stand up in. I managed to stow all my bags in the lower bunk, and tried to figure out how two people could live side by side in that cabin without constantly treading on each other.

The cabin came complete with its own heads (or toilet to you landlubbers), a cupboard-sized bathroom just big enough to take a man, with a pump-action toilet which temperamentally sometimes didn't dispose of its contents quite as well as it should have, and a cold shower. Despite the shock to the system, this was a godsend for getting clean after a day in the salt air. It was a strange experience showering whilst watching the waves pass the porthole a foot away, or trying to complete one's daily constitutionals whilst hanging on for dear life as the boat pitched and rolled.

The other hazards of the forward bunk became clear overnight, when most of our travelling was done, so as to reach a new island by dawn. In November, the season is only just beginning and the sea can still be rough. On several nights I found this to my cost, as each wave would literally lift me bodily out of my bunk and drop me back down again. The whole boat would rattle in these conditions. With all this and the deep throb of the engines from the stern, sleep was at a premium and often curtailed before first light by the dropping of the anchor, the chain being right above my cabin.

Despite all this, I loved the *Cachalote* from the minute I stepped aboard her. I remember sitting in her stern as we sailed out of Baltra, watching the Frigatebirds playing pterodactyl-like in our wake, and thinking how great a feeling it was to *know* with all certainty that the next week was going to be one of the most unforgettable of my life.

The plan on that first day was to sail to the nearby island of North Seymour, home to thousands of breeding seabirds. I was the only hard-core birder on board, and I suspected these islands were an even greater delight to me than to my fellow shipmates.

By mid-afternoon we were moored just off the island, and I immediately fixated on the whirling flocks of Blue Footed Boobys, and the snowy Red Billed Tropicbirds, with their long tail streamers, being chased relentlessly by the piratical Frigatebirds. The more adventurous took an early opportunity to test how cold the water was (very).

After a typically sumptuous lunch, we went through what was to become an everyday routine – apply sunscreen liberally, fill bottle full of water, remember binoculars and camera, don't forget hat – and were soon chugging towards our first landing in the ship's dinghy.

One thing was immediately obvious – you didn't come to the Galapagos for the scenery. North Seymour was typical – a flat volcanic plateau, arid and devoid of vegetation save for some scrub and hundreds of gigantic cacti, most taller than a man.

But what the island did have was also immediately obvious – some incredible wildlife. I knew this was going to be special when I had to step over a young sea lion pup to get off the dinghy. I seemed to mind this far more than he – as with all the birds and animals in the Islands, the relative lack of human contact over the years had meant that they were absurdly tame. Lying in my bunk, listening to the evocative honking of the sea lions each time we made landfall before it was even light, was one of the memories of the Islands that will stay with me forever.

We were no sooner ashore at North Seymour than we were introduced to one of the talismanic animals of the Islands, the land iguana. These brutes could grow to around 4 feet long, and were the subject of much photography as the males sparred with each other amongst the rocky outcrops, for all the world as if though they were putting on a show for the tourists. They looked like extras from *The Lost World*.

Sonia was a mine of information – she told us how the iguanas on each island were subtly different from each other, and led us to a thick patch of low scrub where there was a huge Frigatebird colony. Here the males would sit at about head height, inflating the immense red sacks under their chins to attract passing females, and making their strange clicking and whirring calls. Judging by the number of nests in the colony, they were doing this with some success.

I knew that the bird life would not add huge numbers to my List, but that the majority of the birds I did see would be endemics, which is to say found nowhere else on the globe, and therefore immeasurably valuable to any lister.

The mission for the most ardent birder in the Galapagos was to see all thirteen species of Darwin's Finch. These were all peculiarly adapted to their particular niche in the local ecosystem, and were the central plank in Darwin's theories of evolution – adapt or perish.

However, they were especially boring birds, with spectacularly boring names – Medium Ground Finch, anyone? I decided early on that I would not go out of my way to see the whole collection of Finches (in the end I got eight of the thirteen), but would rather revel in the more extraordinary birds the islands had to offer.

One of my favourites was to be found on most of the islands we visited, the Swallow Tailed Gull – now *that* is how to name a bird. This large gull, with its large bill and long tail, was immediately recognisable by its striking wing pattern, and was usually to be found in the shadows of the cliffs since it was one of the few species of gull which has adapted to a nocturnal existence. Like the many species of Petrel to be found around the islands, this lifestyle had been adopted over the centuries to prevent the ubiquitous Frigatebirds from stealing the Gulls' food as they returned from their fishing expeditions at sea.

You only had to see flocks of the marauding Frigatebirds harassing the elegant Tropicbirds until they regurgitated their catches – caught in mid-air by their tormentors – to see why this technique had been developed, and why the Swallow Tailed Gulls had evolved such a striking wing pattern – surely to be seen more easily at night. It was not for the last time that I marvelled at the adaptability and durability of the natural world, as I sat on the lava beach on North Seymour and watched the sun set of the shimmering ocean, the brisk sea breeze blowing the caps off the waves in the fading sunlight and the sea lions cavorting in the surf only yards away.

As you might expect from a group of islands so far out at sea, their chief ornithological interest was seabirds in general, and in particular the multitude of species of Shearwater and Petrels that abounded in the local waters.

I soon discovered that the prime spot for sea-watching when we were on the move was the stern of the boat, which was covered from the sun by an awning and afforded views both to port and starboard, as well as aft. The only distractions were the smell from the engine room underneath, and when the ship's cook needed to access his store of fresh fruit and veg, which was kept under my seat!

Sea-watching at home invariably meant hours swathed in windproof clothing, wiping the salt from your lenses, trying to keep your eyes from watering and endeavouring to keep your scope on that speck in the ocean before it disappeared – try giving directions to a bird to a fellow birder in an environment without a single topographic feature. It was a particularly idiosyncratic branch of birding, sea-watching, which seemed to appeal most to ruddy, long-haired, taciturn individuals who, one imagined, had spent too many lonely hours gazing at nothing from some desolate headland.

Not so sea-watching here in the Galapagos, however. Here the only hazards were the sunlight reflecting off the ocean, and the smells coming from the galley as cook perfected another of his masterpieces. The birds were sometimes tricky to keep track of as the boat rolled and pitched, but generally seemed to stay with the boat for long periods – just as well given the intricacies of seabird identification.

In this way, the List was swollen by the addition of those mysterious birds that spent the vast bulk of their lives at sea – Audubon's Shearwater, Elliot's Storm, Wedge Rumped, Dark Rumped and Madeiran Petrels, and Waved Albatross all were picked up from my vantage point astern.

With such broken nights and long days, everybody was safely

in bed soon after dinner, but there was ample opportunity to get to know my shipmates. The Brits travelling together had mostly been travelling by coach, south from Quito through Columbia, Brazil, across to Bolivia, and back up through Chile and Peru. They were full of stories of open latrines and freezing glaciers, and of the magic of Machu Picchu. There were times, however, when their stories and in-jokes threatened to swamp the rest of us.

Eric was our solitary German, and did a great job of integrating himself with the others .He worked in a bank in Frankfurt, but had grown up in East Germany before unification and seemed the embodiment of how unification had given the East Germans opportunities they could never have dreamt of before.

I was very jealous of Eric's perfect English and also French, and made up my mind to at least improve my pidgin French when I returned to the UK. I asked Sonia, who spoke decent English as well as Spanish, French and, of course, Italian, what the secret was. She looked at me as if the answer was obvious, then said with much waving of the hands:

'Well first, Russell, you must move to the country you want to learn the language of. And then, of course, you must take a *lover* from that country.'

'Well,' I thought, 'that's going to work wonders for Naaz's French once she lives with JJ in Switzerland then.' An unwelcome intrusion of reality when I least expected it.

Paul, or Paulo as he liked to be called, was a living embodiment of Sonia's theory. Born in Italy and raised in the Home Counties, he had moved back to Italy and was a business consultant based in Milan before undertaking his own adventure. His Italian was near perfect.

After Ecuador, this intensely intellectual but eccentric individual was to rough it through Brazil, before spending a year travelling the hard way across the Pacific (visiting Easter Island, no less, just about the most remote place on earth) and then back home through Australia and India. Inquisitive about others' lifestyles to the point of irritation, I nevertheless liked Paulo immensely, partly because I empathised with anyone who had left a lucrative lifestyle behind to pursue a dream.

Finally, our American couple was an enigma. Both previously married, Ron and Barbara were, in typically American fashion, undertaking this trip together to see whether they were suited as a couple. Ron was a real bohemian, a rabid vegetarian, yoga fanatic and immensely spiritual in his outlook. Barbara was more bookish and genteel, and definitely more in love with him than he seemed to be with her. Both were appalled at Bush's election victory, now confirmed, and mystified by the heavy sarcasm and crude innuendo that passed for the British sense of humour around the dinner table.

These were all people who, although I didn't like them all, I was glad to spend time with, as we were all on our respective journeys of some kind. Despite a very heavy cold, which couldn't have endeared me to them in such a confined space, we got on well and the spirit on board the *Cachalote* was one of the highlights of the trip to the Galapagos.

Over the next few days we visited a number of different islands, each offering something new and different.

On Santa Fe, we took the dinghies out into the shallow bay, and watched as the white tipped reef sharks and golden rays slid effortlessly and silently under the boat. It was mesmerising to glimpse a fin or a tail and try to imagine what the whole animal looked like, even though it was only inches below us. Ashore, the beach was populated by a family of sea lions, most of whom dozed in the sunshine as if we never existed. How many countless thousands had been butchered in this way by our ancestors around the world, I wondered.

It was on Santa Fe that I encountered another Galapagos endemic bird, the Galapagos Hawk. Although a rather nondescript brown raptor, they gathered ten at a time over the boat anchored in the bay and allowed you to walk within a couple of feet of them – a real experience to see such a wild bird at such close quarters.

Feeling rather weak and sorry for myself, with a constantly streaming nose, I dragged myself onto the walk around Espanola, and was very glad I did. It was here that we first saw marine iguanas, the Galapagos being the only place on earth where this species had adapted to a life spent eating seaweed and swimming, and having evolved a gland on its head to expel the salt with an almighty 'sneeze' that I could certainly identify with.

Further inland, we came across colonies of Masked and Blue Footed Boobies, sometimes nesting right on the path so that we had to step over the sitting birds or fluffy nestlings. You didn't need a good camera to get incredible shots of these birds. The local speciality was the Waved Albatross, a near endemic bird that nested on only a few islands. This year the colony had inexplicably failed to breed, and so we were grateful that we were able to find a few birds still socialising amongst the volcanic boulders.

Each island also had its own subspecies of Galapagos Mockingbird, once again accentuating the fact that the Islands were an evolutionary laboratory. These birds had learnt that tourists mean food, and more importantly on these islands, water. On several occasions one of our party would put down his pack to take some photographs, and return to find a Mockingbird rifling through its contents!

The long journey to the main island of Santa Cruz gave Sonia the chance to fill us in on the many threats to the Galapagos Islands. These came chiefly from the growth in immigration from the mainland, which was bound to grow further given Ecuador's economic chaos.

Although some of these people worked in tourism, many more were fishermen, and there was growing conflict between the authorities and the fishermen about access to the Islands and fish quotas. Like the supposed limit on the amount of tourists to the Galapagos national park, each of which brought in $100 in visa payments, the general thought was that these quotas were poorly enforced and routinely exceeded. It was easy, too, to see how a growth in human population would increase the risks of pollution, as was almost demonstrated by an oil spill from a ship a month after we had left.

Santa Cruz was the biggest island in the chain and one of only two (so far) with a significant human population. Our stay here was therefore very different to that on any other island, and some of us took the opportunity to tour a couple of local bars and a nightclub in Puerto Ayora, a surreal activity given how far we were away from true civilization.

Now a month away from home, I felt as if I was travelling for real – how could I be any further from where I was from and what I used to do than by being amongst the fantastic birds and animals in this remote paradise?

The following morning a couple of our party took advantage of the local facilities and went off to scuba dive, whilst the bulk of us explored the island, the only one we were to see with lush vegetation. We trekked inside an extinct volcano in the interior highlands, now covered in thick forest, where I spotted an exquisite Vermillion Flycatcher and an equally gaudy Galapagos Dove.

But the real highlight was still to come, as we visited a local farm. There amongst the cattle and green meadows were several muddy pools, and to our utter amazement many of them were refuges for humongous Galapagos giant tortoises.

These animals had migrated from the highlands to forage lower down, and seemed a throwback to another age – nowhere else had the environment allowed this species to grow so large. To see them wallow contentedly in the muddy water, slowly chewing on the succulent grasses, was a wonder, as if we were privileged to see a creature from antiquity. Indeed, such is the longevity of these creatures that it is entirely possible that some of the very largest would have been alive over160 years before, when Darwin himself was a visitor!

In a world where long ago progress had seen the demise of most of the larger species on earth, it was hard to dispel the notion that these were living on borrowed time as well. Watching them flinch at every loud noise we humans made left me thinking that we were an unwelcome intrusion, although I would not have missed these extraordinary animals for the world.

The sheer fragility of their existence was brought home by a visit to the Darwin station, a research facility near Puerto Ayora. Each island had its own species of tortoise ('Galapagos' is one of the old Spanish words for turtle, or tortoise) and many were represented at the station, but none got more attention than Lonesome George, the last remaining specimen of his species. Given his sexual incompatibility with other species, a race was doomed to die with him.

A final spree to buy postcards and T-shirts and we were at sea once again, this time alighting on the north coast of Santa Cruz, where I added to the List with three Greater Flamingos feeding in a shallow lagoon and a lone Galapagos Martin.

The afternoon was spent moored at Sombrero Chino, where I

felt a little awkward about not joining the rest snorkelling off the local beach – being *below* the waterline was never my idea of fun, especially with a blocked nose, but I had promised myself I would try this before I went back to the mainland – the question was, would I live up to it?

Instead I did a little sunbathing, feeling slightly guilty having read of the floods back home, and wrote some Christmas postcards – a futile gesture as it turned out, for as far as I know every single one of these is still languishing in the 'post box' at Baltra airport – who says Ecuadorians don't have a sense of humour? In any case, nobody ever saw I word I wrote that afternoon.

The weather was turning more sultry, and that night most of us laid out on deck looking at the stars, until the heavy swell drove us below decks once again. There is nothing that is guaranteed to make you wistful than a spot of stargazing. There, right on the Equator, with no light pollution at all, the heavens positively shone with the myriad of stars, planets and galaxies. I wondered about James at home, wishing he were safe – I'd managed to get past his mother just once to talk to him on the phone in the month I'd been away, and had begun to dream about him incessantly.

I asked myself whether Naaz was looking at these same stars, and told her how much I loved her. And I pondered on my journey, the risks I had taken and the adventures to come. I really loved being in such a remote and unique environment, and truly felt free of all my worries back home at last. I banished all thoughts about my future to the very back of my mind.

Genovese was probably my favourite island, for the sole reason that it was great for birds. One side of the island was dominated by a massive Petrel colony, and the walk along the cliffs amidst clouds of wheeling birds was one of my birding highlights of the Trip. There must have been hundreds of thousands of birds all told.

This island was also the only one to have all three species of Booby nesting – Red Footed were added to the usual Masked and Blue Foots – and I was fascinated by how they used trees to nest in, given that their bright red, webbed feet were so obviously ill-designed for the job. Discovering a Short Eared Owl roosting in a crag below the path suitably rounded off the morning.

Later, on Darwin's Beach, I finally plucked up the courage to don a mask, snorkel and flippers, but the waves were so strong I could barely stand up, and even just putting my face underwater and trying to breath made the mask fill up with seawater. I gave up after a few goes, feeling a fraud – there can't be too many visitors to the Galapagos who haven't snorkelled. I reconciled this to myself later – after all, it was *my* Trip, and if I didn't snorkel, I didn't really care. Besides, there were no birds under water. Or, so I thought.

On our final island, Bartolome, we undertook a major trek across a fresh lava field – that is to say, only a few hundred years old. It looked like a huge vat of tar had boiled over and thickly coated everything for miles. We then moved on for a last swim from a nearby beach. It was whilst wading in the surf, watching a little jealously the others snorkelling off shore, that I realised there *was* one species of bird that could be encountered underwater.

The first I saw of it was a glimpse of a shape just underneath the surf – at first I thought it was another sea lion, which were happy to swim right up to any swimmer. The next thing I knew a couple of Galapagos Penguins had swum right up to me, and as I took one step forward, one actually swam *between* my legs!

These endearing birds had not been seen before on our voyage, and I was delighted to make their acquaintance at such close quarters. Originally having arrived from somewhere further south like Chile, this species had stayed, nourished by the fish stocks present due to the cold water currents, and evolved into a separate species, thousands of miles north of the latitudes you would usually expect to find Penguins. Where else but the Galapagos would you find Frigatebirds above you and Penguins below?

As the sun set we all climbed up the side of another extinct volcano and, despite the cinders in the shoes and the lungs about to burst, it was worth it to watch the sun setting over the bay, with our little *Cachalote* looking like a toy boat far below.

Up at dawn on our last morning, we had time for a quick boat trip to the mangroves to see mating green turtles, and were lucky enough to spot several lurking amongst the roots like gigantic terrapins. We then hurriedly packed and came ashore at Baltra once more. We handsomely tipped Sonia and the crew (another

Third World custom you ignore at your peril) and bade our reluctant goodbyes.

It was somehow fitting that, as I looked back at the *Cachalote* one last time, with a lump in my throat, watching the Frigatebirds circling like vultures overhead, I almost jumped out of my skin as I nearly stepped on a complaining sea lion that had chosen the wharf for its afternoon nap.

At the airport, I jealously watched the next set of tourists 'de-plane', few of whom could have anticipated the richness of their experiences to come. I wonder if any of them realised that one bad oil spill, or even worse an *El Nino* which permanently disrupted the Humboldt Current, could turn this wildlife paradise into a desert in a tiny fraction of the time it had taken for it to evolve.

My week in the islands had only yielded another fifty-two species, and yet twenty-two of these were to be found nowhere else on earth, and most had been seen within touching distance. It had been a birding expedition you could never forget.

And so, my first Big Adventure was over. It had been a humbling experience, and made me appreciate just how big the world out there is. It also wet my appetite for more adventures to come. However much I looked forward to a warm shower again, I would never ever forget the *Cachalote* and the magical islands of the Galapagos.

BIRD COUNT: 225

5 Gordon the Gecko and Ed the Attorney: The Andes and the Amazon, Ecuador, November – December

Having bid our final farewells over a Thai meal in Quito, the passengers aboard the good ship *Cachalote* went their separate ways. I had a couple of days on my hands in the city before my next adventure began, in the Amazon basin far to the east.

I didn't fancy another day stuck in the capital, even though it had come alive in our absence with the 'Viva Quito!' festival, a week-long shindig celebrating not independence from Spain as one might expect, but the conquest of the Indians by the Spanish conquistadors over 400 years before. Nothing like rubbing their noses in it, I mused.

The celebrations seemed to largely consist of each trade or district renting a truck, and driving around the city centre with a brass band and lots of very drunken people aboard. Routinely a few of these dropped off and were run over, or were maimed by fireworks, but I guess you can't blame the people from wanting to close down their businesses and party for a while, especially given the state of the economy.

I hurriedly arranged with Rodrigo, the local guide, that he would take me up to see the local mountains, and especially the largest of the volcanoes hereabouts, Cotapaxi. Although I was resigned to another day of headaches, having gone from sea level to over 2 miles high in twenty-four hours, this trip also had the considerable attraction of lots of extra birds. In areas such as this, altitude, not latitude, was the biggest dynamic for bird distribution and I looked forward to some exotic additions to the List. And with money so short, Rodrigo wasn't going to complain about earning a few extra dollars, even if it meant working on a holiday.

First there was the not insignificant matter of flirting with Naazlin, who I hadn't been in touch with for over a week and who was about to get even further away by visiting her mother in Kenya for a fortnight. It was weird having an intimate chat with her in the small hours in Quito, whilst she was at work on the other side of the world.

I didn't know it at the time, but international telephone calls are the only expensive item in Ecuador – no wonder the Internet cafés are so popular. A bill for $800 at the end of my stay in Quito certainly made me question whether all the effort, heartache and expense was really worth it, but already our ability to stay in close contact whatever the circumstances had become a feature of each new destination, a way of reliving my adventures, of sharing them and another challenge to be overcome. I wasn't going to give up now.

Yet another early start saw Rodrigo, our driver and I on the road up into the Andes at first light. The Quito rush hour was just under way, with long queues to get the ubiquitous buses into the city for those that still had work. Some 20 miles or so out of the city, the roads began to get distinctly provincial, and the long drive began to assume a familiar routine, consisting of suicidal overtaking manoeuvres interspersed with zigzagging to avoid the potholes. If the altitude didn't make me feel queasy, then the driving was surely going to.

Around mid-morning we stopped at a couple of Indian towns to visit their markets, for me an unwelcome touristy distraction, but fascinating nonetheless. I was surprised at the lack of hassle, seeing as there were only a handful of westerners present. The main goods on sale seemed to be wool garments, usually in very bright reds and greens, that the locals seemed to prize highly, especially as long socks, worn outside their trousers or tights.

There were the usual collection of pan pipes, cheap jewellery, fake alpaca sweaters and other tourist clutter, whilst other stalls had steaming vats containing the entrails of unspecified animals, or sometimes even an entire one – usually a guinea pig. Were it not for the fact that these vats smelt greasy and disgusting, I might have been tempted myself.

As usual though it was the people themselves that were the most fascinating. Most were less than 5 feet tall, and all seemed to favour the short-brimmed hats and llama-wool ponchos of the region. Their faces were deeply lined and weather beaten, regardless of age, their teeth rotten and brown, caused by the unrefined sugar they all chewed in order to provide extra energy at such high altitudes. For me every exertion left me gasping for breath, and I marvelled that these stocky resilient natives could actually farm the mountain slopes in these conditions.

I was impatient to see more of the mountain range, and we were soon off again, to cross the range and descend to a national park on the other side.

Several people on board the boat in the Galapagos had told me of Cotapaxi, the highest active volcano in the world, and how you could climb halfway up it, and get great views if the weather was good. The problem was, this was far too energetic for me in the rarefied atmosphere of over 5500 metres (they had already acclimatised in Peru beforehand), and in most cases the majestic mountains were covered in cloud. Driving through the park gates, I could already see that the cloud cover was low, but we had no choice but to press on.

Despite being designated a national park, in true Latin American style any attempt to actually *conserve* the area was merely notional. The *paramo*, or high altitude steppe, had largely been given over to forestry or had been degraded by goat herders and their animals. It made for a depressing journey up the mountain road towards the plateau at the top, where the cloud was so low that the mountains were out of view, and there didn't seem to be any bird life at all as we ate our picnic lunch in a brooding silence.

As we got higher, however, things improved, and a sudden break in the weather smiled on us, as for half an hour the cloud partially lifted and there was Cotapaxi in all her glory, a lava cone capped in snow and looking like a primeval ice cream cornet. It was the only colour in a barren and washed out lunar landscape.

The mood lifted along with the cloud as we parked on the plateau, and Rodrigo and I went on our bird walk around one of the mountain lakes. At first sight this looked pretty lifeless, but away from the road, and amidst the eerie stillness and quiet, birds began to show themselves under Rodrigo's prompting – for without a field guide I was reliant on his experience, and his tatty *Birds of*

Columbia back in the bus for identification.

Fortunately for me, he knew all the official names, and not just the colloquial South American ones. And by God, what names some of these had – first to show themselves was a splendid little dusky finch, the Plumbeous Sierra Finch, shortly followed by a Bar Winged Cinclodes. The habitat seemed to be especially suited to thrushes, and within half an hour we had logged Yellow-Legged and Solitary Thrushes, plus a Sooty Robin, one of several species that superficially looks like our Blackbird back home.

I was also pleased to mop up some of the altitude specialities, like Andean Gull, Quail and Lapwing, the latter spectacularly noisy as it tried to warn us away from its nest site. However, the best bird of the trip I actually managed to find myself.

Rodrigo took us over to a patch of red flowers, this being the favoured feeding area of the Andean Hillstar – unbelievably a tiny hummingbird uniquely adapted to a life at altitude, which hibernates each night, in effect, closing down its systems to conserve energy and heat. In the very thin air at nearly 6000 metres above sea level, it had to flap its wings faster than any bird alive to stay in the air.

It was with great relief then that, just as Rodrigo was about to give up the search, I caught sight of the pale belly and black breast band of a Hillstar on the other side of the small valley – probably the most unusual bird since I left London, and all the sweeter as I spotted it first.

By the time we turned for home in the mid-afternoon, the early start and the altitude had got the better of me, and I was all set for a long doze on the way back to Quito. But Rodrigo had other ideas.

For about an hour he talked about everyday life for him and his young family in Quito, how he and is wife both had university educations but worked in tourism because that was now more lucrative than many of the professions, especially as people like me paid in cash, and in dollars. He also explained how he had lost all his savings when the banks suddenly shut their doors before the recent crisis, and how the new president, brought to power in a peaceful army coup only a few months before, had promised a long-term payback plan. It was one that nobody believed would deliver the lost savings back to their rightful owners – after all, the previous president was now living in exile in Venezuela, enjoying a lavish lifestyle courtesy of his embezzled millions.

Rodrigo seemed to be sanguine about this tragedy, and I admired his fortitude. He had vowed not to join the mass exodus of the middle classes from Ecuador, saying that his country was going to need people like him and his wife to get out of this mess. I couldn't imagine anyone back home taking this situation on the chin like Rodrigo, and accepting part of the responsibility for solving the country's ills.

He also was an expert on Inca history (Quito being an ancient Inca town at one time) and enthralled me with tales of the Spanish invasion and the Inca resistance. One story has stuck in my mind ever since:

> It seems that when the Spanish finally overcame the Inca capital, they held the Emperor to ransom, expecting his Empire to be rich in jewels and precious metals, and needing to send home a large fortune to pay for the expedition.
>
> The Inca Empire was indeed exceedingly large and fabulously wealthy, and a famous Inca general spent two years accumulating this wealth from the vast interior of the Empire, in order to pay the ransom.
>
> On his way to hand over the fortune to the Spanish, the general learnt that the Spanish had gone back on their word, and executed the Emperor, a God as far as his people were concerned. In a furious rage, he decreed that all the riches must be hidden in a remote spot that no man would ever be able to find, so that at least the Spanish would not prosper from their duplicity.
>
> The vast fortune was hidden deep within a cave network. Every soldier who helped deposit the treasure there was executed, to keep the location secret. Then the Inca general and his fellow nobles all committed suicide.
>
> The search for the Inca gold has been an obsession, rather like El Dorado ever since for modern man. There have been numerous expeditions to find it, but all have ended in failure and have been dogged by a succession of catastrophes, in

much the same way as those visiting Tutankhamun's tomb are reputed to have suffered. The last expedition was some twenty odd years before, a Franco-German affair from which no one came back alive owing to the 'curse of the Incas'. A rescue mission discovered only a video camera, complete with some film. This apparently was under lock and key in the USA, and its contents top secret.

Rodrigo paused, and I assumed the story was over, but with a very solemn look he added a peculiarly personal twist to this tale.

In his travels as a tourist guide, he had spent some time in the mountains, and become good friends with a local witch doctor, or shaman, of one of the villages. He had won her trust, and one night she recounted the story of a dream she had had many years before:

In the dream, she had been awoken by three Inca warriors, who requested her help. Though afraid, she agreed, and was blindfolded, strapped to a spare horse, and taken on a three-day ride to an unknown location. Although blindfolded, she could tell that at first she was high in a mountain pass, and then crossing a scorching desert, then finally in a deep dank forest, before reaching a large cave.

There she was untied, and the blindfold removed. She spent some time caring for an Inca chief who was desperately sick. Her potions helped him to recover, and eventually she was thanked and led back, tied and blindfolded as before, to her home on the other side of the mountains.

To my cynical Western mind, this sounded at first like pure hokum. Then Rodrigo explained that she had been telling the story of the pass, the desert, the forest and the cave in her village for many years, ever since the dream happened.

However, it was only very recently that rumours had begun to circulate in Ecuador of what was on those mysterious videotapes found in the mountains and now kept secretly in the USA.

Apparently the Ecuadorian government had finally been allowed to see the tapes. They showed the expedition climbing a mountain pass, and then trekking across a baking desert, before descending into a cool rainforest. The last footage was of the explorers entering a large cave complex. None of them were seen again, and no bodies were ever found. So how did the shaman know all of these secret details, and years before her own government?

Was this a fairytale for the tourists, or some thing more spiritual at work here high in the roof of the world? I do not know, but it was damn good yarn and, yes, Rodrigo did get a larger tip than usual that night. It might have been hokum after all, but he seemed to believe it and I had come to not only like but also trust him. For that moment at least, I believed it too.

My second free day in Quito was rather more prosaic. The local airline, Equatoriana, had gone bust, and I hurriedly had to reorganise my flights so as to make the connection through to Buenos Aires in a week's time. It was here, shuttling between bureaucrats at several travel companies before finally getting the new ticket I needed, that I really learnt that speaking Spanish was a massive advantage in South America.

I had enough time to visit Quito's Old Town, and found that this part of the city did indeed have a lot of character. I explored the honeycombed streets and took in all the faded grandeur and flaking colonialism. A statue of an angel high on a nearby hillside dominated the area but, typically, I was advised not to visit it as it had become a hotbed for pickpockets and muggers.

In truth I found myself kicking my heels, awaiting my departure back into the wilderness, this time further east. I love to visit cities when travelling abroad, for that is where all the people are and, therefore, where you can take the pulse of the country you are in and soak up the local flavour. But it is the countryside where I really belong, and as I sat and ate a solitary meal of pork, beans and popcorn, I was literally counting the hours until the next journey, and my adventure into the Amazon.

After a restless night being kept awake by the firecrackers and all-night parties of 'Viva Quito!', I met up with my new travelling companion at the airport next morning. Mr Edward F. O'Connor was talking in what sounded like fluent Spanish to our very pretty tour guide when I arrived. Despite this (I was jealous on at least two counts already) I grew to like Ed immensely as soon as we started to chat on the short flight out across the mountains, heading for the steamy rainforest and our first stop, the oil town of Coca.

Ed was an attorney in Newport Beach in Los Angeles, but his Californian drawl and carefully manicured hair and beard were the only typically American things about him. He had a nice line in self-deprecating humour, accentuated by the dog's breakfast of an election that his country had just participated in – in fact, as a committed Democrat, Ed seldom let an hour pass without raging against a system that had put a president in power with less than half the popular vote.

I recognised a kindred spirit, in that Ed was single and young at heart like me, or so we both kidded ourselves. Ed despised his fellow Americans who never left their own shores. After travelling many times to Asia and Europe on business, he had hoarded his vacation time and air miles to grab the opportunity to experience the rainforest. In his own cocooned little way he was doing the same as me – breaking loose to see the world whilst it was all still there.

He had done his research, knew a little of what was to come, bought himself a flash new digital video recorder and sharpened up his already strong Spanish. I was to spend the next five days with Ed, and was already looking forward to it.

But most of all, I couldn't wait to renew my acquaintance with the rainforest. I'd been to Costa Rica the previous year, and the sheer profusion of life was staggering, from insects to plants to birds, all of which seemed to be painted in exotic colours or have mesmerising calls. Not for nothing was South America called the 'Bird Continent' – I had seen over 400 species in just two weeks in Costa Rica and had high hopes of making a big dent in the List here in Ecuador.

The rainforest always seems so primeval, as if once the entire world had been this hothouse of natural wonders, before we came along to tame it. And the fact that it is disappearing fast is another good reason to soak it all up, a privilege our grandchildren may never have.

This was perfectly summed up as we flew into Coca. Twenty years before this was all virgin rainforest, but then oil was discovered and suddenly Coca grew like mould on bread to service the industry.

This was about as near to a frontier town as I'd ever seen anywhere. The airstrip – I wouldn't have used the term 'airport' to describe the runway and tiny building – was right next to the town itself. We negotiated our way through dozens of ragamuffin children, the most enterprising of which had learnt a little English and were desperate to carry our bags. We met our guide, Gus, who said it was OK for them to help, and we duly forked out a dollar each only to discover that they had carried our bags all of 20 yards and dumped them in the back of the dusty flatbed Ford on the kerb. This was the vehicle that was to take us to the Napa River, to our boat to La Selva Lodge deep in the Amazon basin.

We met a group of Americans coming back from La Selva, en route to Quito. They had had a wonderful experience and all seemed to have turned native in their short space of time in the jungle – bandanas, thongs, shorts, sandals and a profusion of brightly coloured shirts that only Americans seem to wear. It whetted my appetite even more for what was to come.

Our bumpy ride along the dirt road to the river took us through what passed for a town centre – ramshackle shops selling engine parts or cheap clothing, tin-hut cafés serving oil men, and teenage prostitutes lurking on street corners looking to do the same for a handful of petro-dollars.

It was a relief to get out of the truck and into our long-prowed boat, which was to take us two hours up river, out towards the the Columbian and Peruvian borders. It was with no regret at all as we left Coca, pausing only to fill the rest of the boat with crates of Coca-Cola, proving that even here on the edge of civilisation, this is the world's ubiquitous drink. Perhaps even the town was named after the stuff.

The Napa was far from the majestic waterway I'd expected. In the humidity of an Amazonian squall, it seemed muddy and languid. For the first half an hour, we chugged through the drizzle, past tumbledown farm shacks on the riverbanks, wooden canoes full of naked children, and ominously, whirring helicopters ferrying materials to the road workers deep in the jungle. How long before all of the Napa looks like Coca, I wondered?

Gradually the weather improved, and we left most signs of civilisation behind, occasionally passing through some busy little township on either side of the river, or a happily waving family fishing from their jetty on the riverbank.

Ed, however, missed all this, being glued to his pulp fiction paperback. He may have wanted to travel, but I began to see that like many Americans he actually wanted the rainforest to be brought to him – he had no intentions of actually interacting with it, or involving himself with it – he wanted to consume it in carefully choreographed, easily digestible portions. Me, I wanted to be *part* of it.

I was glad that our lodge was so far from humanity, so that it felt we were actually in the wilderness. By the time we had left the launch tied to a sandbank, and descended into our canoe for another hour's paddle to the lodge, I had begun to smell and feel the forest around me.

There didn't seem to be much bird life along the river, but as soon as we took the canoe along a creek, which led to the lake and the island that housed La Selva Lodge, everything changed. Black Capped Donacoblus chased each other through the undergrowth, Hoatzins, with their ridiculous long necks and ostentatious crests, squawked at our intrusion and red howler monkeys danced from tree to tree, seeming to follow our passage upstream. A Black Caracaras soared overhead. Gus's T-shirt stated 'In the Rainforest anything is possible' and I was already starting to believe it.

We bumped into another canoe carrying an immensely corpulent American birder, also staying at the lodge. He had been out since dawn and his exhausted demeanour and damp combats reminded me of the scene in *Apocalypse Now*, when Martin Sheen battles through the jungle to find Colonel Kurtz – our sweat-soaked birder looked like he hadn't seen another white man for years, never mind days.

The lodge itself looked a complete dream. The main building was reached via a long wooden walkway from the jetty, and was a big thatched hut that acted as bar, lounge, library and dining room in one. Straight out of the canoe, Gus organised some very welcome drinks and fresh pineapple and, amidst the screech of the Yellow Rumped Caciques nesting outside the window and the colony of extraordinarily noisy Crested Oropendolas above us, he explained our itinerary.

Basically, La Selva organised any jungle activity you wanted, complete with personal guide – one of the advantages of being one of only three guests. There were usually two walks daily, one at dawn, another in the late afternoon, with the middle part of the day too hot for birds or animals and given over to resting up (and in Ed's case, reading pulp fiction). Wellingtons were mandatory, since the rainforest was always wet, rain or not. Mosquitoes, amazingly, did not seem to be a major problem, but at night long-sleeved shirts and trousers were wise investments.

I was quickly allocated to Jose, an Indian and the resident bird expert. In broken English he suggested we go for an immediate bird walk, as it was already mid-afternoon. I was in my element and, after dumping my bag, we were soon on the boardwalk, which connected the island to the mainland, and birding furiously.

Most of the English Jose did know turned out to be bird names but he was a master at bird calls and, as we trekked through the forest paths, he would pause, look up into the canopy and then issue some exotic call with a combination of lips, tongue and fingers. More often than not, this would attract the bird to within spotting distance, at which point Jose would drag me with a 'Quick, Quick, Come' into a position where I could see the bird. The most difficult part was deciphering the name of the bird from his heavily accented description.

In this way I accumulated Purple Throated Fruit Crow, an audibly challenged Screaming Piha, a Plain Throated Antwren, a gorgeous Blue Crowned Manakin and several types of the dozens of species of Flycatchers in these parts within an hour, and we then turned for home as darkness fell and the mosquitoes began to buzz.

The lodge really came into its own at night. There was a generator, but this only provided enough electricity for the kitchen and for the electric lights in the hut toilets – all the rest of the lighting, in the huts and the lodge was by gas lamp.

After a meal almost on a par with those on board the *Cachalote*, I returned tired but exhilarated to my hut, where the staff, who outnumbered us by about three to one, had left a couple of gas lamps burning. With the help of this and my torch I was just about able to see enough to undress, but the lack of light was the least of my worries. Spiders kept me awake that night.

Or, to be more accurate, the fear of spiders. I'd had this phobia ever since I was very small, and can remember what seemed like a huge creature running down my face one day as I played behind the sofa at home. My father and brother both hate them also, to this day, and it was my mother or my wife who had to remove them from the bath all through my adult life.

And now here I was in the realm of the spider. I had learnt on previous expeditions in Africa and Central America in the past how to take what I thought were sensible precautions – always turn shoes and boots upside down after taking them off, and shake them vigorously before putting them back on. Always check the bed thoroughly before getting into it. Always wear something on your feet when walking to the toilet, and always check the toilet, sink and shower for vermin before using it. And, touch wood, I had yet to encounter one of the beasts in the tropics.

However, the signs for further success weren't good. The walls of my wooden hut were made of small wooden poles bound together, and in many places there were gaps big enough to push my fingers through. This was presumably why there was more wildlife *inside* the hut than outside – a sizeable green gecko, which I called Gordon, lived in my bathroom and often watched me shower after a long day in the forest. In my roof, there were a colony of bats that would have kept me awake if I had not been so knackered every day. It's surprising how tiring just walking in 35 °C of heat and 100 of humidity can be. And every morning, there was my own private collection of insects nestling in the toilet bowl for me to examine.

However, this didn't stop me enjoying the nights almost as much as the days at La Selva. Apart from the rustles and scuttling of the animal neighbours in my hut, the sounds outside in the forest proper were truly magical.

After nightfall, the forest became the realm of insects and amphibians, and they provided a natural concerto so extraordinary that you didn't want to sleep and miss part of it, although the lodge actually sold tapes of it for you to take home if you did. I especially liked the gentle metallic sounds of the small tree frogs, which, although no bigger than your thumbnail if you were to find one, were usually a stupendous shade of indigo or scarlet. I was in heaven.

The next three days were a complete orgy of birding, plus a physical ordeal I was not expecting. On the first day, Jose decided to test my mettle, and we canoed over to the other side of the river and embarked on a long bush walk.

This was hard going, especially with the terrain continually broken by hills and streams, and with me still trying to acclimatise to the heat after spending some time at altitude. I spent much of the time trying to keep up with Jose, only managing to do this when he stopped for birds; he didn't seem to feel the need to pause to let me rest in the mean time. By mid-morning I was sweating heavily, particularly in my long trousers and wellies, but Jose looked as fresh as a daisy. By noon I was wondering if he would ever stop for our picnic lunch, and in the end I had to ask him to call a halt so we could both eat, drink copious amounts of water and, most importantly, rest. It was still idyllic, however, as we sat in a glade to eat our pork, beans and rice, and watched the butterflies land on our rucksacks.

And, of course, despite the heavy going, we saw tons of birds, a total of 64 on that first full day alone, or about 25 per cent of the birds for the entire trip so far, and all in a single day.

Rainforest birding is a peculiar pastime. Many of the hundreds of species in the forest are brightly coloured, others have loud and obvious calls, but almost all are hard to find unless you have a Jose to call them in for you. This is because half the species live in the canopy, where you might spend half an hour gazing straight up to see them flit around in the tallest trees, sometimes hundreds of feet up.

Most of the rest are the opposite; they skulk deep within cover, often on or near the ground and, as if this wasn't hard enough, they are mainly black or brown in colour. After a day birding in the rainforest, you know that if you have a sore neck from looking straight up, and a bad back from crouching right down, then you have had a good day.

That first day was a very good day. Before the trek, we visited a salt lick by boat, an area of the jungle where there is easy access to deposits of minerals in the earth, which parrots in particular need to help digest their intake of unripened fruit. The forest in this area at dawn was a cacophony of noise, as Mealy, Blue Headed and Yellow Crowned Parrots flew around in vast flocks, accompanied by Dusky Headed and Cobalt Winged Parrotlets.

Then on the walk itself, highlights included a pair of smart Blue and Yellow Macaws, five species of Hummingbird, another five of Woodpecker, and Bay Headed, Green and Gold and Swallow Tanagers, the most brightly coloured of all species in the rainforest and my all-time favourites.

By mid-afternoon I was keen to get out of the forest, and at last we emerged into a little bay where someone had left a boat for us. Jose took us up river, to a steamy little island, where we scrambled ashore with some difficulty. The target here was the Umbrella Bird, an immense Amazonian crow-like species which a platform of ruffled feathers protecting its face.

After a sweaty hour of cutting our way through the undergrowth, and with Jose imitating the calls, we finally identified a bird high up in a mature tree, and with patience, and Jose's extraordinary ability to sense the bird's next move, I eventually got a glimpse of the whole bird, a really good tick.

Nor was the day over. After dinner, we scrambled into the canoes complete with flashlights and liberal amounts of mosquito repellent, and searched for caiman, the local species of crocodile. After an hour all we had to show for our endeavours was a cloud of ghostly Sand Coloured Nighthawks, disturbed just before dark from their roosting trees, and later on several fiendish-looking greater bulldog bats hunting low over the water.

Eventually we saw our quarry, two neon-red eyes picked up in the flashlight, at water level. We inched closer to the red dots, and got to within a couple of yards before the caiman disappeared under the water with a loud, startling splash.

Gus had been hoping to catch and haul one of these animals into the canoe so we could take a closer look, as he had managed to do about a week before, but to Ed and I this seemed utter madness. We were keener to try and find an ocelot or even a puma, but Gus explained that even scientists living in the jungle studying them seldom saw these. He himself could count the numbers he had seen on the fingers of one hand, he said. We were surprised he had any fingers or hands at all given his recent encounter with the caiman.

So we couldn't see everything the rainforest had to offer, but that first day in La Selva was one of my best birding days ever, and one of the most memorable of the entire Trip.

It soon became clear that Ed was not the adventurer he claimed to be. With his sardonic humour and heavy sense of irony, he was content to read in his hammock for much of the day or 'rage against the machine' over dinner, but he actually managed to miss much of the splendour of La Selva during his stay there. He would manage one walk or boat trip a day and then just chill. I couldn't help but think 'For heavens sake, Ed, you get one hard-earned vacation a year, go out and enjoy it, see the world!'

Ed laughed this off with his usual good humour, and in fact had the last laugh. On one of his sporadic forays from the safety of his hammock, it was he and Gus who managed to see an immense anaconda slither off a tree branch and into the water, when Jose and I were still deep in the forest.

Not that I didn't value my own time in the hammock – its surprising how comfortable they can be, dozing, wafting in the breeze, listening to the mysterious bird sounds from the forest beyond – but I threw myself into every walk and boat trip I could get Jose to take me on, although I did decline when he playfully suggested another all-day trek.

We spent a fruitful couple of mornings on a platform built high in the canopy of a gigantic tree, the perfect way to see forest birds without neck ache. At that height it was possible to look out over a sea of green and pinpoint the birds as they flew at dawn from their roosts to their favourite feeding trees.

In this way, we picked out huge birds like the top-heavy White Throated Toucan, and its smaller cousin the Many Banded Aracari – tropical birds have such extraordinary names, as befits their appearance. With such a gamut of species, the English language tends to become inadequate and, typically, a three- or even four-word name is not unusual.

It was up here in the canopy platform that I really begun to appreciate the paintbox colours of the rainforest birds, so bright as to be almost unnatural. There was the navy blue and lemon Golden Bellied Chlorophonia, the piebald White Necked Puffbird, the deep indigo of the Short Billed Honeycreeper, the snazzy Green and Rufous Kingfisher, the irrepressibly gaudy Bay Headed Tanager and the incredible Paradise Tanager, a tiny bird covered in every colour of the rainbow, like some psychedelic sparrow. But my personal favourite was a Spangled Cotinga, seen on top of a tree some quarter of a mile away, yet still a wonderful confection, a small dove with a coat of soft sky blue and an iridescent purple throat.

Meanwhile, our walks deep into the forest produced the other kind of rainforest bird, the dark mysterious ones which you had to call in and then grub through the undergrowth (forgetting about spiders as you did so) to try and get a glimpse of – come on down, the Plumbeous Antbird (grey), the Rufous Capped Anthrush (brown) and the Spot Winged Antshrike (black and white).

This whole family of antbirds were named as such due to their amazing habit of following parties of army ants around the forest. These formidable insects, about an inch long or so, roam the forests in columns comprising hundreds of thousands of ants, overpowering insects, frogs, birds, lizards and even small mammals who might get in their way. The Antbirds gather on the sidelines, and wait for the creatures that are desperate to get out of the swarm's way to pass by – classic ambush tactics. A whole food chain for an entire birding infrastructure is dependent on the habits of these tiny but deadly insects.

In addition, we came across several blue morphs; huge blue butterflies the size of a dinner plate, floating effortlessly through the forest on several occasions.

On our last afternoon, Jose took me in the boat to an island in the middle of the Napa, where the habitat was more savannah-like and more new species beckoned. We were able to add birds like Yellow Browed Sparrow, Chestnut Bellied Seedeater and the impressive Ladder Tailed Nightjar to my by now burgeoning List, before a plague of midges, desperate to enter every orifice, defeated even the redoubtable Jose and we were driven back to the sanctuary of the lodge.

Our last meal was over too quickly, and Ed and I duly did our Father Christmas act a couple of weeks early by dispensing our largesse to everyone and his dog in the form of dollar bills. It is to Jose's credit that, having missed the handout the previous evening, he was too proud to come and ask for his (considerable) tip on the boat the following morning – I had to get the boat to return especially to thank him in the time honoured way. Other travel guides from around the world, take note!

There was still one small drama to unfold before our last night spent listening to the 'Amazonian Night Orchestra' for the final time however – the discovery of a large tarantula in the roof of the bar.

Ed was instantly on the bar and up in the roof, video camera in hand, but I was only to pleased to be watching this monster squeeze itself to safety through the slats in the roof from a distance. Given my fears on that first night, I guess I would have settled for just this fleeting encounter beforehand.

The journey back along the Napa the following morning was punctuated by several stops to load people and possessions going to market, and by the time we got to Coca it was full to the gunwales. The journey back to civilisation was rather surreal, as less than six hours after leaving something akin to the lost world we were back in dusty old Quito again.

Despite Ed's idiosyncrasies (or may be because of them) we got on well, and the following day, our last in Ecuador, was spent being typical tourists. We took a cab to the monument right on the Equatorial line (after which Ecuador was named by the Spanish), a rather impressive tower sitting astride the magic yellow meridian, and flanked on all sides by the barren mountains of the Andes.

It occurred to me that it was strange that here I was, with one foot in each hemisphere, thousands of miles from home, yet back in the UK I had never been down the road to Greenwich to do with the same with longitude as I was doing here with latitude.

Ed proved himself to be the commensurate American after all, talking loudly to everyone within earshot, secretly glad I think to be back in civilisation again. Whilst I stood and tried to let the significance of standing on the Equator sink in, he went off to haggle over the price of alpaca sweaters.

We had a last dinner together, and in a desperate attempt to join in with the final night of 'Viva Quito!' we cruised a few bars, only to be reminded that we were twice the age of almost everyone we met. It was with some resignation and not a little sadness that we bade our goodbyes.

I left Quito airport for the final time, itching to find a city that suited me better but knowing that the natural experiences in Ecuador might never be equalled.

I picked up my connection for Buenos Aires in Santiago, Chile. Here I briefly considered flirting with the Swedish girl next to me, and then thought better of it when she told me of her conversion as a born-again Christian. Not much chance of a wild night out *there*, I thought.

Instead, I spent the time thinking of all those great birding experiences. Whilst the Galapagos had contributed a small, high-quality list, the richness of the rainforest at La Selva had more than made up for it – I had seen more species in four days there than in four *weeks* in North America!

That birding injection had boosted the List considerably, and now, with only a quarter of the trip completed, I was over 40 per cent of the way towards the magic 1000 bird total. Surely it was achievable now, I thought, surely . . . perhaps I should even revise it up to 1200? What about 1500? If only I knew then how difficult it was to become.

I looked out of the plane window. The majesty of the rugged, snow-capped Andes was giving way to the flat, stark, barren Pampas and, beyond, the vibrant city of Buenos Aires. I was exchanging one of the poorest, most unspoilt countries in South America for one of the most civilised and sophisticated.

A couple of hours later, we were banking over the River Plate and making our approach into Buenos Aires airport. I was about to find out if the differences were worth having.

BIRD COUNT: 403

6 Welsh Whales: Patagonia and Iguaçu, Argentina, December

Buenos Aires looked and felt like the big city it was. The ride in my limo took me along the freeway from the airport and into the city centre, and on this steamy Friday evening, as the sun set, half the population seemed to be barbecuing on the roadsides in their vests and shorts.

As we got nearer to the city proper, more freeways began to appear, and then skyscrapers and shopping malls. This was all much more like the USA than South America, and certainly nothing like Quito.

Having checked into my hotel and gone exploring, more contrasts with Ecuador began to emerge. The centre of BA (as everyone called it) was actually quite European in atmosphere, big wide boulevards reminiscent of Paris, an opera house like a poor version of the *Scala* in Milan, a huge obelisk in the centre similar to Cleopatra's Needle in London, and shopping centres like you might find in Madrid or Barcelona.

The people were noticeably different too, and this was clear even in the few hours I had to kill before bed and an early start on the way south in the morning. I had read that over a third of the population were of Italian extraction, and sitting outside a café, sipping a beer and people-watching, the passers-by did have a certain old-world patrician grandeur about them. There seemed very few Indians here. Other countries in South America seemingly regarded the Argentinians in the way some Europeans regard us Brits – reserved and a little superior – and as Latin Americans who identify more with their ancestors in Spain and Italy than with their neighbours in the hotter countries to the north.

For a single man however, BA was a welcome return to form after Quito. In the sultry night air it was impossible to miss the stream of leggy Latin lovelies that paraded past, their noses in the air as if not to notice that every red-blooded male within a hundred metres was looking at the length of their stilettos and the shortness of their skirts.

I had little time to dwell on all this, however, and by dawn the following morning I was back at the airport to catch my plane south to Trelew and Patagonia.

Patagonia, a vast expanse of grassland that occupies the southern half of Argentina, was my first stop largely because of the Valdes Peninsula on its northern coastline. This was one of the few reliable places to see the southern right whale in the world, and a must-see for anyone as keen on the animals as I had become.

The southern right whale was a close cousin of the northern right whale (surprising, I know), the first whale to be hunted in any numbers in Europe, and named after its ability to float when harpooned – it was the 'right' whale to catch for the whalers.

As such, it was also the first whale to be nearly exterminated, first in the northern hemisphere and then in the south. The small numbers left south of the Equator bred close inshore in a handful of locations, but none of them were so accessible or as famous as at Peninsula Valdes. Add to this the large colonies of elephant seals and sea lions (and who can forget those great shots in Attenborough's *Life on Earth*, where the baby sea lions are taken from the beach by marauding killer whales – all shot right here in Valdes) and I was expecting another wildlife spectacular. In fact, after three flights and three countries in less than twenty-four hours, I'd have settled for anything that didn't involve sitting still and eating airline food.

However, the first sight of Patagonia from Trelew airport was not inspiring. As we drove to the hotel, Patagonia looked very flat, very grassy and very very windy. It was, however, positively interesting compared to my first sight of Trelew, one of the few sizeable towns in this part of the country. This was one of those foreign towns that looked permanently closed, as though as a foreigner you were the only person who had not been allowed to share some dark secret that had driven everybody else away.

However, a new country always got the adrenaline flowing and before long I was met in the lobby by my driver for the next few days, Harald. Harald had been born in Patagonia, the son of an

Argentinian father and an English mother. This had left him with a good command of the English language and a rather unfortunate allegiance to Leicester City.

Sharing his battered Cortina on that first day were a Jewish couple from Buenos Aires, out for a day trip to see the sites. Neither spoke much English, but both talked incessantly, with much rolling of eyes and gesticulation of hands, about their children, the prices in BA, how poor the food in the hotel was – in fact anything but the landscape they had come to see.

Like a true professional, Harald pretended to be interested, whilst pulling me aside at every opportunity to apologise for their intrusion, tell me that all folk from the big city were a pain in the backside anyway and reassure me of his personal attention as soon as he was shot of them. The day was getting more surreal by the moment.

An hour later we were at Punta Tombo, home to half a million Magellanic Penguins. Every tourist to Patagonia comes here, but amidst the assembled throng, I managed to lose Mr and Mrs Jewish Couple and find some space where I could take it all in.

I stood atop a small cliff and looked down on the beach. Below were thousands of Penguins, resting after feeding, clambering up the beach to their burrows or meandering down to the water. In the shallows I could see little posses of birds swimming, for all the world looking like they were flying through the water on their stubby little wings.

Secreted around the beach and the dunes were thousands of burrows, most occupied by a sitting bird and a few containing a fluffy grey chick. The Penguins went about their business with their beaks in the air as if they knew they were gawped at by hundred of tourists. In fact, there was more than a passing resemblance in this respect to some of those Latin lovelies in BA – or perhaps I had just spent too much time on my own recently?

The Penguins mooched about, greeting one another, fighting one another, tending their young, occasionally looking rather curiously as a tourist loomed closer to take a photo – I swear I even saw one of them smile, they were so human.

Around the beach, I was able to add my first Argentinian birds – on the beach were Guanay Cormorants, Buff Necked Ibis, and the local speciality, the Chubut Steamer Duck. An inspection of the tideline yielded a Snowy Sheathbill, a white pigeon-like bird from the Antarctic, and a reminder of exactly how far I was away from home.

We drove slowly back to Trelew, and I began to appreciate the bleak landscape and, eventually, even to ignore the yattering in my ear from Mrs Jewish Couple. Outside, the bumping Cortina scattered Elegant Crested Tinamou in every direction, and passed a hunting Chimango Caracara about every mile or so.

Harald stopped when we spotted a family party of Patagonian hares – actually more like a huge mutant guinea pig – and the local version of the llama, the guanaco. In this way we made our way in our own personal cloud of dust back into a local town, and it was here that the day got even stranger.

The Welsh had arrived in Patagonia some 150 years before, where, despite the complete absence of mountains, they had stayed to raise millions of sheep on the inhospitable grasslands. This part of Patagonia was liberally scattered with towns named after their homeland, and even today many of the shop fronts were testimony to their ancestors – I'll never forget a sign for one of the local solicitors, called 'Evans, Gonzales and Williams'.

This heritage was reinforced again at our next stop. We drove across the dusty flatlands and then quite suddenly entered a verdant valley, all oaks and farms, and suddenly we could have been in Pontypool rather than Patagonia.

As if this wasn't strange enough, we pulled into a little courtyard and there was a pristine, whitewashed, carefully manicured Welsh Tea Shop. This little cottage in the middle of nowhere was festooned with pictures of the Princess of Wales, including some of her visiting said establishment on a visit to Argentina some years before. She was seemingly held in even higher esteem in the middle of Patagonia than she was in Middle England.

However, this was just about the only English thing in the place – certainly the staff, like all the local people, only spoke Spanish and Welsh. The day got weirder still when we sat down for afternoon tea, and we attacked the mountain of crust-less sandwiches, jam sponges and laver bread to the sound of a traditional Welsh baritone male voice choir – *singing in Spanish*. I was beginning to feel like a character from Alice in Wonderland. Mr and Mrs Jewish Couple urged me to eat and eat more, as if I was their long lost son. If there had been chicken soup on the menu, I was willing to bet they would have forced me to eat it.

Which would have been a blessing actually, judging by the standard of the restaurants in Trelew. I had finally got rid of my well meaning but irritating companions, and that evening managed to find a bookstore selling a book on Agentinian birds, although judging by its scarcity, birding was not a popular pastime in these parts. As it was, it was all in Spanish, although in the small print it did at least have the English names. And, like everything in my new country, it seemed to cost a fortune.

Despite it now being early evening in Trelew, the place had scarcely woken up. There seemed to be little traffic to speak of and only a few stores had people in them. It had the overall impression of an English provincial town on a sultry Sunday afternoon.

I had been told that the Argentinians love to eat, and in fact they are the biggest meat-eating nation on earth. What I had not been told, but what Harald put me straight on the following day, was that your average Argentinian family has dinner at home and seldom eats out. This would explain why only the supermarkets seemed busy, and why, despite searching for a couple of hours in every tawdry back street in Trelew, I managed to find the grand total of two restaurants.

As darkness fell, I walked into one with a huge gaucho outside, and sat down near to the door, so I could see the sweating cook turning the huge hunks of meat on a massive grill in the window. Finally an apologetic waiter came over to speak to me in English that wasn't so much broken as smashed to smithereens. He sheepishly explained that his restaurant, representing half of the culinary output of the entire town, didn't even open for another couple of hours.

I would have killed even for some more laver bread right then, but managed to hold on and eventually enjoyed a steak the size of a house. I'm not sure which was bloodier, the meat or the spectacularly violent Argentinean football on the television as I ate.

I had Harald to myself for the remainder of my stay, and he turned out to be a rather odd cove. A thickset bear of a man, with a paunch that was a monument to Patagonian lamb, he had sallow skin, as if the constant wind had bleached the entire colour from his face, and deep-set, very dark eyes that seemed to convey a huge weariness of his world.

Later, Harald told me that he was recently married for the third time, and also that his eldest son had been killed in one of the many car crashes on Patagonia's long, straight but rocky roads that seemed to encourage suicidal driving. Little wonder that he wore the face of a man who had seen everything.

I think Harald was more pleased that anyone that there were just the two of us now in his ancient charabanc. He loved to talk, about how the people in Buenos Aires were awful (it seems the inhabitants of capital cities are despised in the provinces the world over – a classic have and have-not situation), the tumbling price of lamb, the merits of respective tourists from foreign countries, and his new diet. Meanwhile we rumbled across the vast landscape, where the sky seemed so big and brooding, and thunderstorms ambled across the plains – a rarity this, as it rains here on only about ten days each year.

This morning was the highlight of my time on Valdes, a drive down to Puerto Ballenas to see the whales. As I waited for the previous boatload to come in from the breeding grounds further out in the bay, I scanned the water and could periodically see the surface broken by a small black tail, the right whale being one of those photogenic species that loved to 'fluke', that is, flap its tail above the surface. The adrenaline rose, and once again I began to appreciate just how lucky I was to be here on the other side of the world and sharing the water with such an eminent creature.

Soon it was our turn and we climbed aboard the boat and sped out into the bay. At first all there was to see were gangs of Kelp Gulls and South American Terns feeding on the 'boils' of fish, and a succession of dead penguins, blighted by some unknown virus that year all along the coast.

Our captain however had seen some flukes a long way off, and before long we were slowing down to drift in a calm patch of water where the whale was last seen. Great clear patches on the surface of the sea showed where the whale had flexed with its tail underwater to move away, and our captain followed these 'footprints' until suddenly the whale majestically broke surface only a few yards ahead.

Like most whales, viewed from the surface the animal seemed to be a dark indeterminate mass, with flippers and flukes breaking the surface at unexpected angles and times. Those unfamiliar with whales tend to forget that it's eyes and mouth lie low in its head, and so they will seldom see the 'face' they are expecting. However, the whale has to periodically surface to 'blow' or breathe, so there are regular opportunities to study it. The sound of a whale blow, like a prehistoric steam engine, is not to be missed and seldom forgotten. And, just like their cousins in Canada, the smell of the whale's breath was like a slap in the face with a rancid haddock.

The right whales are particularly gruesome, even in a family of animals not blessed with beauty, by dint of having barnacles encrusting their head and snout, the grey and orange of these contrasting with the black and white of the animal itself. However, this was far from my thoughts as our first whale performed for the camera and rubbed her chin along the side of our boat at point blank range.

It soon became obvious that there were dozens of whales in Puerto Ballenas (which is named after the whales). Many of them were mothers with calves, and we watched enthralled as the huge submarine-like adults torpedoed past the boat, followed by the much smaller calves, some almost completely white, but all keeping within flipper distance of their mothers. They seemed just as curious about the boats as we were about them. Seeing these family groups together was to appreciate their strong social hierarchies, which can include aunties, older children and babysitters as well as mothers within the same groups. It also emphasised, if ever it needed it, how barbaric the slaughter of these fantastic animals is.

Exhilarated and humbled by my latest whale encounter, I felt as though I was walking on air as I walked back to Harald on the beach. How often he had seen that look on a visitor's face, I thought, and yet never once felt the urge to go on the boat and see for himself the wonders I had just experienced.

Instead, he drove me right around the peninsula, stopping for a meal of local barbecued lamb and visiting a local beach, which was notable for having colonies of both sea lions and elephant seals strewn along it for several miles. They may have shared the same beachfront, but there the similarity stopped. The sea lions were typically active, cavorting in the waves, chasing each other across the sand, honking furiously like a gaggle of demented geese.

The seals, on the other hand, had made a science out of just lying around. Occasionally, one would saunter further along the line of corpsed companions, the fat rippling along its flanks as it tried to move in an environment that it was not made for. Both species were sleek at sea, and somewhat clumsy on land, but only the seal had made it an excuse for lying around and doing nothing.

I stood on the cliff top trying hard not to let the omnipresent wind carry me over the edge, and looked out over the sea. I could see the ugly Southern Giant Petrels quartering the ocean, and the much more attractive and demure Dolphin Gulls hoping to pick up scraps from a sea lion's meal. I tried to imagine a pod of killer whales making an assault on the beach to steal away young sea-lion pups, but knew that those famous sequences had taken months to capture on film . . . still one can dream.

The other bird I was desperate to see was the Lesser Rhea (the local version of the Ostrich), a bird as distinctive as its home on the stony flatlands. Harald and I spent hours driving along dirt roads expecting to see one of these curios, but we were nearly home before one popped up on a small rise as we turned a corner, and scuttled away across the plains, its long neck and tiny head somehow seeming almost unconnected to its huge feet and legs.

That night was spent in Puerto Pyramides, a tiny beach town near to the whale-watching centre, where I intended to go out again in the boat the following morning. I eventually negotiated some superb seafood pasta and local beer, but not before once again discovering that an absence of Spanish in these parts made life a lot harder. One thing I did discover about the language was that, judging by the volume used by all my Argentinian neighbours late at night in my hotel, it didn't matter what you said as long as you said it *loud*.

The wind that was so characteristic of Patagonia also cancelled the whaling trip the next morning, and it was a rather forlorn tourist that Harald picked up later, having kicked his heels gazing out to sea for a couple of hours, when he had expected to be *on* it instead.

Harald tried to cheer me up by going to a number of birding sites he had used with other birdy tourists, but all seemed to be as quiet as the grave. The weather turned ominous, and the wind got stronger as we headed back towards Trelew, the clouds as dark as my mood.

Harald said he had one last place to visit, if he could still find it. On the wrong side of town, amidst the shanties that you wouldn't want to be near after nightfall, was a huge rubbish dump. It was also a major economic centre for the local inhabitants judging by the dozens of them, mostly scruffy little urchins, who were combing through the piles of rubbish as we drove in.

Harald looked uneasy at the prospect of parking here, but I persuaded him to carry on and as we rounded a bend in the track I was very glad I did. There was a large natural lake, part of which contained lots of rubber tyres and general flotsam from Trelew. But there, against this very uninspiring backdrop, were hundreds and hundreds of birds – ducks, geese, swans and cormorants.

The ability of nature to adapt to the presence of man (or die out trying) has always surprised me, but not often as much as coming across the best birds I'd seen in Argentina in the middle of the town rubbish dump. I got out, much to Harald's consternation, braced myself against the car as protection from the gale, and happily drunk in the sight in front of me.

There were probably over 500 wildfowl on the lake, from South American staples like Southern Wigeon and White Winged Coot, through less common species such as Brown Pintail and Cinnamon Teal, to elegant Coscoroba and Black Necked Swans. Feeding amongst them were a few White Faced Ibis, and over sixty fuchsia-coloured Chilean Flamingos.

Suddenly my mood lifted with this unexpected bonus, and Harald seemed genuinely pleased that one of his ideas had borne fruit. I didn't even mind having to spend the afternoon back in the proverbial one-horse town of Trelew (or should that be 'two-restaurant town'?).

To brighten my mood further, I finally got through to James

back in the UK, and delighted in hearing about his news from school and his football team. I then hunted down the local Internet café, and caught up on all the news from Naaz. I don't know whether it was because I was on a high or not, but sitting there reading her email, sent from Nairobi where she was still on holiday, made me realise that the world had actually become a very small place. That was the first time I resolved to see whether a surprise trip back to see her was practical – mad though it seemed at the time.

As an Englishman in Argentina, I naturally worried a little about whether I would see any anti-English feeling as a result of the Falklands war. Although it took place nearly twenty years before, I knew that the Argentinians were a fiercely nationalistic nation and that the defeat still rankled with them.

Harald had never brought it up and so far I had not even seen any graffiti, so when on the drive to the airport the following day we came across a vivid mural courtesy of the Argentinian Air Force depicting the bombing of British ships and underscored with an emphatic 'Las Malvinas es Argentinos' I was taken a little aback. Something about the chuckle from Harald told me he had deliberately saved this pleasure for last.

I made my connection in Buenos Aires with minutes to spare, and was somewhat surprised to find myself on the border with Brazil, and in the middle of the rainforest again, by mid-afternoon. The steamy jungle and waterfalls of Iguaçu can scarcely have been a bigger contrast to the arid and windy steppe of Patagonia.

Flying into Iguaçu, you could see the spray from the falls, rising up over the flat forest like steam from an enormous kettle. By the time I had reached the hotel at the falls proper, I could see that this detour to the north just to experience them was going to be well worth it.

This area of Argentina is unique, in that it is the only part of the country that is tropical rainforest. The heat and humidity as you emerge from your cocooned capsule of an aircraft takes your breath away, the beads of sweat instantly like bullets on your brow. This made Ecuador feel positively temperate.

The town of Iguaçu was quite some size, but it was only there at all because of the falls, which like their counterparts in Niagara and Victoria were magnets for tourists and honeymooners alike.

Not that the falls themselves were any less impressive because of this; they stretched on both sides of the border with Brazil for nearly 2 miles, with literally dozens of cataracts when the rains were heavy. They were so much more impressive than Niagara as a result and, because on the Argentine side you could access many of the falls by boardwalk or boat, they also had the edge over Victoria, where it is difficult to see the entire vista except from the air.

I checked into the palatial hotel near the falls, one of the most expensive I've ever stayed in; it is the only one near enough to walk right up to the action and therefore charges accordingly. Desperate to see the water close up, I immediately booked a trip for that afternoon and within the hour was travelling through the forest in the back of a truck to board our boat.

The drama was heightened by the instructions as we boarded: waterproof all your cameras, don ponchos and lifejackets, remove all wristwatches. After a few minutes on the river, I began to see why. Ahead of us were the first serious of falls, including one called 'El Diablo', which really was a devil to get close to, such was the boiling torrent at the foot. Yet this was where we were going, actually *under* the falls in our boat!

Desperately trying to use the camera without getting it soaked, part blinded by the sheet of spray that covered my glasses from hundreds of metres away, and already wet through to my skin, I gazed up at the awesome curtain of water as it got nearer. At point-blank range, the captain accelerated the boat into and out of the cascading water before we had time to scream. An exhilarating experience, and one which had us all wanting 'More! More!' like little children.

Back on dry land, I used the rest of the afternoon to scout the local trails and take in all the majesty of the falls (and dry out). Every turn seemed to yield another impressive vista and I soon realised that I would need much more film than I'd brought.

The falls carried on for series after series, some framed by trees, others with their own personal rainbows, others still punctuated by rocky outcrops in their midst – look hard and you could almost convince yourself that you could see them being eroded, such was the colossal volume of water. You stood at the lip of a waterfall and watched thousands of gallons disappear over the edge in what seemed like seconds, marvelling at where it had all come from. Even in the *rain*forest, how could there be so much water around? Could there be any left at all further upstream?

The other big bonus about being back in the neo-tropics again of course was the bird life. Some of the species were the same as in Ecuador, but such was the profusion and variety in the rainforest that many of the bird families had new species down here, further south.

I wandered around the trails, my clothes gently steaming in the sultry heat. The most obvious new bird was the Great Dusky Swift, thousands of which fed on the insects hatching in the surrounding waterways, and nesting actually *behind* the falls on the cliff face. To stand and watch a fall for any length of time was to marvel at these birds as they flew straight at the falling water, kamikaze-like, and emerged on the other side, where if you looked carefully enough you could see dozens of them clinging to the wet rock face.

The birds here were just as colourful as I'd come to expect – the Green -Headed Tanager, Scarlet-Rumped Cacique, Toco Toucan and Chestnut Eared Aracari, to name just a few on that first afternoon. Iguaçu was a potent cocktail of fantastic scenery and spectacular birds.

I was less keen on the hotel, however, for any number of reasons. First, it had just decked itself out in Christmas decorations, and it was a surreal experience to sit in 35 °C heat and watch the angels sparkle on the tree. Also, most of the guests were there in big family groups and this accentuated the fact that I was by myself, the first time I had felt lonely since Key West. Additionally, it was the big impersonal hotel that you might expect in such a location, and a bit of a rude shock compared with my cosy log cabin at La Selva.

There was one further surreal experience to come at Iguaçu. At breakfast I met a guy who seemed to be by himself, a clean-cut all-American kid named Dan Murphy. We got chatting, and he told me he was awaiting the arrival of his girlfriend who was currently trekking in Peru. As you do, we got chatting about our jobs. I recited the by now well-rehearsed story of how I'd given up everything at home at forty-two to fulfil an ambition to see the world and its wildlife.

Dan gave me the usual slightly disbelieving stare, and then explained that he was an attorney. God, not another, I thought! What was weirder was that he then told me that his practice was in Newport Beach, Los Angeles. I was sure this was the same town in which a certain American lawyer I met in Quito had had his practice, and so I said: 'You're not going to believe this Dan, but not only are you the second attorney from the States I've met in a week, but the other one also had a practice in Newport Beach.'

Dan looked a bit surprised, and asked me where I'd met the other guy. 'In the rainforest in Ecuador; you might know him, his name is . . .'

'Ed O'Connor!' Dan said with a cry of triumph. I felt my heart jump and the hairs on the back of my neck rise.

'Do you know him?' I asked incredulously.

'Know him? *I sit right next to him!*' Dan said.

So it was official, the world really was getting smaller after all. Here I was, seven days and several thousand miles from where I'd last seen Ed, having breakfast with the guy he shared a desk with. Unbelievable.

That morning I'd planned to go to the head of the falls, a horseshoe crescent called the Witches Cauldron, where there was water and spray on all sides. It was a classic honeymooners spot and I found myself thinking of Naaz, wishing she could see this with me even though she was on the other side of the world.

There had been plenty of overnight rain, and the falls were heavier than ever, great spouting columns of water gushing from every crack and fissure. Despite the continuing drizzle, it was an awesome site. We then took a bus ride over the border into Brazil, where I got an unexpected extra stamp on my passport, although my tour guide had to smile very sweetly for them to do it. It was a lot wetter than the Copacabana would have been, but at least I'd made it into Brazil. Now all I needed was a beach babe in a thong and I would be well happy.

Most of the great waterfalls seem to be on national borders, and you usually get the chance therefore to visit both countries. Such is the economic disparity around the world, that very often the difference between the two countries is quite marked. For example, Zimbabwe is made to look prosperous by the poverty across the Zambezi in Zambia, and even at Niagara the Canadian side is quite orderly and well mannered compared with the tacky brashness of the US side.

Igauçu was no different. The Argentinian side of the border had good roads and a prosperous town, whilst the Brazilian side had dilapidated roads and tumbledown shanties. And, I was told, if you took the trip down river and crossed into Paraguay from Brazil, you would see a corresponding change for the worse yet again.

One thing that Brazil did have, however, was some great views of the Argentine Falls from the other side of the river. This was worth the two hours we spent in Brazil, the shortest visit to any country on the Trip, and with the shortest Bird List – Roadside Hawk and Pale Breasted Thrush being the only new additions to my total.

Back in Argentina, however, the bird count grew as I explored the trails and grounds of the hotel, but most of the best birds were found at the end of the day, as I relaxed naked, trying to cool down on my balcony with the coldest of beers. I counted the birds as they went to roost in the forest right outside my door: Screaming Cowbird, Chalk Browed Mockingbird, Rufous Bellied Thrush and tiny birds like the blue and yellow Bananaquit, or the stunning Versicoloured Emerald hummingbird.

At this point I had not spoken to Naaz more than once in a fortnight, the longest time we had been out of contact since we met six months before. Was it really just six months? It seemed I'd known her all my life.

I took advantage of my newly civilised surroundings to try and call her in Nairobi. Perversely, I also couldn't resist making the sacrifice – at $12 *per minute*, this had to be the world's most expensive phone call ever.

Telecommunications between the Argentine rainforest and Kenya are not what they might be, and it took me a dozen attempts before we finally managed a five-minute conversation, with us both screaming down the receiver to try and be heard. Still, money well spent I thought.

My last morning in Igauçu was spent with Daniel, a local guide to whom I paid a king's ransom for the privilege of a personal bird walk around the local forests. As if to accentuate the divide between birding at La Selva and here at Iguaçu, Daniel possessed none of Jose's skills in terms of calling in birds – he didn't need to. Around his waist he wore a belt containing four CDs of bird calls he had personally recorded in the forest over many years. Around his neck he wore a CD player, with a small microphone on each hip. In his pocket he carried a laminated card detailing which bird's calls were on which CD. When he wanted to attract a bird out from the forest or down from a tree there was a brief delay, a blur of technology and suddenly the call was being transmitted from Daniel's nether regions – ingenious.

Like all good guides, he knew the right sites for all the unusual species, although we kept stopping due to torrential downpours to shelter in his jeep. We found a Slaty Breasted Wood Rail, Surucua Trogon, Red Breasted Toucan, Magpie Tanager and White Bearded Manakin, as well as sundry species of Woodcreeper and Tyrranulet, all classic rainforest species. We also saw groups of coati mundi, the local long-nosed, long-tailed relative of the racoon, two black-nosed caimans in the river (where I renewed my acquaintance with the mosquito despite the weather), and the agouti, something of a cross between a giant guinea pig and a rabbit.

Hearing a call he recognised in the forest, Daniel set off through the bush in pursuit, me struggling on behind, and we finally emerged into a small clearing. The bird continued to call, moving around the perimeter of the clearing without us seeing it at all. Daniel, however, was patient and finally he pinpointed the calling bird and managed to put me on to it. In the depth of the thicket sat a Streaked Flycatcher, a bird we had to work satisfyingly hard to get and one which had a special importance as the final one on my List from the Americas.

Back in Buenos Aires I had a free day before I left the Americas for good. With Christmas only some ten days away, I bought gifts for my friend's boys in Sydney and a couple of extras to send home to James. The afternoon was spent on a pilgrimage to a part of the city known as La Boca, or literally 'the mouth' of the River Plate.

My journey out to La Boca showed me that Buenos Aires was far from all wide boulevards, tango lessons and glitzy shopping malls. The suburbs could be run-down, and none more so than La Boca, traditionally the home of immigrant dockworkers, and now not much better off.

To raise themselves above their poverty, the inhabitants of La Boca had decorated their tin shacks and tumbledown houses in lurid reds, yellows and blues, and had erected caricatures of famous local citizens in an irreverent tribute to the better half of Buenos Aires. There were only a few square blocks of this still remaining, a target for all tourists to the city who gawped at the multi-coloured shanties complete with Eva Peron and Diego Maradona leaning out of the upstairs windows.

However, the pilgrimage for me was to the home stadium of Boca Juniors, Argentine soccer champions and, since a win over Real Madrid the previous week, the official *World* Soccer Club champions. The streets were still bedecked in Boca's blue and yellow colours as were every local restaurant, one of which I stopped in to eat the traditional Argentine lunch of meat, meat . . . and more meat.

Unfortunately the Boca's stadium so dwarfed the surrounding streets that it was impossible to get a decent view of it from the outside, and I found no way of getting inside. I had see enough Argentinian soccer on television to know that the fans of Boca had not had their passion for the game diminished by their success. In fact, to the contrary. In the UK we seem to be losing the link between the clubs and their local community, an umbilical cord so obviously still unbroken between the footballers and inhabitants of La Boca, in a classic rags to riches story.

And so, after nearly eight weeks, I boarded a plane in Buenos Aires and left the Americas behind. They had been a curious mixture of civilised banality and fabulous natural wonders, but as time had gone on I had adapted to travelling by myself and was now ready for a complete change of scene. I readied myself for the challenges and adventures of a new continent across the Pacific in Australia, not to mention the thought-provoking experience of the time lords actually 'stealing' a day of my life as we crossed the International Dateline.

Most importantly, I had almost cracked the 500 bird total, and with two-thirds of the Trip remaining. Mathematically, I was a month ahead of schedule. I knew that the 'Bird Continent' was always going to boost the total in a way that no other venue on the entire trip was able to, but surely now, with a new continent's birds to come and the target already looking within reach, I could afford to take my foot off the gas a little?

I hoped so. I had spent almost two months travelling constantly, never more than three nights anywhere except for on board the *Cachalote*. I was staying with friends in Sydney, across the other side of the world, and was looking forward to spending some time kicking back over the Christmas and New Year period, where I knew I would be desperately missing Naaz and James. Just take some time off from the List for a while, Russell, I said to myself . . . nothing can go wrong now. Or could it?

BIRD COUNT: 493

7 More Money than Sense: Australia, Malaysia (and a Secret Trip), December – January

I had always loved Sydney. There was something about the exotic mix of the place – the English atmosphere, Californian weather, the Aussie 'can-do' mentality and a Western infrastructure in an Asian context that made it an inspiring place to visit.

It had everything I was looking for in a city – an impossibly scenic harbour, an eclectic mix of cultures, very fine eating and drinking, a generous share of wonderful women (the mix of Aussie beer and warm weather had produced a particularly voluptuous version of the species, it seemed) and some good friends with whom to enjoy it all. Plus, of course, a whole new series of birds to go on the List.

Not that birding was top of mind as my friend Simon drove me from the airport and across the grandiose Harbour Bridge to his home on the North Shore. Although I had only been away some eight weeks, I had squeezed an awful lot into this and it already felt like eight months that I'd been away. The gamut of birds in South America had left my appetite sated for a while and, with Christmas almost upon us, I told myself that I deserved to relax and enjoy the weather. It was midsummer in Sydney and it promised the first sustained period of seriously good weather since I left the UK, currently knee-deep in floods.

It was also great to see Simon and his wife Cate again, two friendly faces instead of the usual array of acquaintances I had been making on my way around the Americas. Simon is a Brit who emigrated to Australia with his Aussie wife Cate in the late 1980s, having met her in the media industry, where most of us had met our girlfriends. They lived for a time near Sue and myself, and we had spent some great nights out together before they emigrated, and then two fantastic holidays with them in Sydney after they moved.

Simon was an all-round sporty type with a penchant for motorbikes and wind surfing rather than football and drinking, by his own admission an advocate of the hedonistic lifestyle. Cate was as petite and demure as Simon was tall and boisterous, always more likely to be found in a designer clothes shop than a bar. I loved Simon's ridiculous sense of humour, Cate's tolerance of his absurd compulsions, and admired the way they had made a big success of their move. Simon was one of the most respected operators on the Australian media scene, and Cate was busy raising their two impossibly handsome, sun-kissed sons. I was dead jealous, as in 1989 I had gone for a number of fruitless interviews in a bid to get into the Australian media market myself.

Simon and Cate embodied the wealthy Aussie middle classes, with their multiple cars, swimming pools, trips to Europe and yacht races in the harbour. They lived their life to the full, eating outside by the pool for six months of the year, taking the boys to the beach after school, taking full advantage of the great restaurants nearby. It was a fantastic environment and I envied their lifestyle and even their lush domesticity. Yes, for me, Sydney had everything.

I spent some time playing cricket (very badly, especially with severe jet lag), swimming in the pool and eating one of Simon's immense barbecues. A third of the way through the trip exactly, I thought, and now definitely a long way from home in every sense.

Which was fine except that Naaz was still at home. I had begun to really miss her again in recent weeks, and I told Simon and Cate the whole story. Somewhat typically, Simon was intrigued as to just what an 'Indian Princess' was really like, and whether she really was 'full of Eastern promise'. Cate, on the other hand, as a good Catholic, looked at the whole thing with a curious and benign disapproval.

I had hatched a plan to dash home and surprise Naaz in London just after Christmas, and both of them looked at me as though I had already had too much sun when I told them that, along with my Christmas shopping, I also had to buy a return flight to chilly, dark, sodden old England just to spend a few days with someone else's girlfriend. They didn't understand that I would gladly travel

AROUND THE WORLD WITH 1000 BIRDS

twenty-four hours and 12,000 miles for just one kiss from the women that would not leave my mind no matter how far I travelled. I felt lucky that I was still young enough at heart to feel like this about anyone.

For the next few days I reacquainted myself with my favourite city – taking yet more photos of the impossibly charismatic harbour, with its imposing bridge and opera house vying with each other across Circular Quay to see who could attract most attention from the many Christmas tourists around. I used up film after film on the two of them, first from the harbour side and then from a ferry across the harbour, getting suitably sunburnt in the hot Australian sunshine for my trouble.

Although I thought I would take a rest from birds whilst I recharged the batteries, it was soon obvious that I was going to picking up new species without even trying. Not only was the fauna in Australia largely unique, offering new species almost at every turn, but Sydney was also one of those cities where the wildlife seemed to inhabit the city quite happily.

Consequently, I found Silver Gulls and Australian White Ibis around the harbour, with Wedge Tailed Shearwaters skimming the waves every time I crossed it. In the Botanical gardens next door, there were lots of unfamiliar birds like Pied Currawong and Magpie Larks, but the Australian Pelican circling high over the Opera House had a very special significance – the five hundredth bird of the Trip – exactly halfway there.

The obligatory Kookaburra laughed hysterically at my bowling in the local park near Simon's house, whilst his garden produced technicolour Rainbow Lorikeets whizzing dramatically through the trees, and other common suburban species such as King Parrots, Red Whiskered Bulbuls, Common Mynahs and Welcome Swallows. And, to my great surprise, a neighbour called me over on a boiling hot Christmas Day afternoon to see three spectacularly ugly Tawny Frogmouths roosting without a care just feet above the lunch table in his own garden.

It seemed in Sydney I was never going to be far from new birds, although even they could not compete with the nightly spectacle of the squadrons of flying foxes making their ghostly way in the twilight from their roosts to feed on trees in the city parks. In the dim light these creatures looked for all the world like a cross between pterodactyls and Lancaster bombers.

All of this allowed me to relax and forget about the progress of the List whilst I explored the city. During the day I bought Christmas gifts for Simon and his family, in the evening we ate and drank wonderfully well at home or in the many fine restaurants on the North Shore, after which I would go to bed for a long evening email session using Simon's computer.

Then came a restless night interspersed as it was every couple of hours by sexy phone calls from Naaz, brightening up her winter's day at the office and taking full advantage of the fact that I was finally in a country where my mobile phone worked.

I also used Simon's email connection to finalise arrangements for my surprise return to London. I was pleased but surprised to discover that Naaz's boss was willing to help in this conspiracy, and together we concocted a bogus meeting for her at which I, and not a new client, would be the lunch partner. I also managed to arrange a romantic weekend at a hotel in Midhurst, and begun to pray that I would have the nerve to go through with it and that she would be as delighted and receptive as I hoped. It was a big risk, made bigger by the complete absence of flight availability back to the UK early in the New Year.

Once the trap was set I knew I had to go through with it, and decided that, as I was only ever going to do this once in my life, I would do it in some style. By the following morning I had booked a First Class return ticket to London on Malaysian Airlines, with a three-day stopover on the way back in Kuala Lumpur. At least if the Nazz thing was a complete disaster, I could take some comfort in the adventure of a luxurious couple of days spent in airline opulence, and a visit to an unscheduled new country – or at least that is what I managed to convince myself of as I handed over the cash. To make me feel even guiltier, I was flying on James's birthday, and knew there would be no time to see him too. I reasoned that the Naaz opportunity was a complete one-off, and that I would have many occasions to make it up to my son. One day he would be a man also, and be able to understand what was going on in my head, as well as my heart.

With the logistics out of the way, it was time to focus on the birds again before Christmas arrived. Simon organised a thrash in the local pub for some ex-pats in my honour, and there one of his mates, as it turned out a manic birder who used to run his own trips locally, gave me a couple of tips as to where to go. The advice was welcome; after five days of chilling in Sydney I had started to run out of new birds and now the List was beginning to look sorry for itself.

The next day saw me in a rented car driving south towards the Royal National Park, some 20 miles outside the city. The weather was steadily getting hotter by the day, and my journey across Sydney in the rush hour and then south meant that I arrived mid-morning, just as bird life would be becoming quieter.

The Royal was truly stunning, a mixture of wooded ravines, heath land and coastal beaches all bathed in that light that Australia seems to have, a special sunlight that automatically seems to accentuate the colours in everything. Like many things in Australia, however, there was cruelty within the beauty – exactly a year later, the park was virtually burnt to the ground by a series of bush fires that came to within a few miles of Sydney itself.

I managed to dash off around twenty new species before the temperature rose into the 90s (°F) in mid-afternoon and both I and the birds gave up. This was time aplenty, however, to discover the myriad species of Honeyeaters, nectar-drinking birds but about as far removed from hummingbirds as you could find – big, bold, noisy and colourful, like a true Australian bird should be.

A little searching provided the List with a small flock of Variegated Fairy Wrens, the males chasing the females with their tails pointing skyward and illuminating the bush with flashes of metallic blue.

However, the best bird of the day, and a speciality of the park, was the Superb Lyrebird. You might think that only a race as cocky as the Australians would consider such an epithet as this, except that they were just being honest as usual. Here was bird reminiscent of a peacock, with a huge plumed tail, and what it lost in colour, being a dowdy grey and brown on the body, it certainly made up for in its calls.

Despite its overly ornate appendage, the bird was not easy to see, and the first thing you knew about a nearby Lyrebird was the extraordinary collection of clucks, rattles and screeches emanating from a nearby bush, interspersed with mimicry of other bird calls and human noises. I actually checked my phone once before I realised the Lyrebird was doing a great impersonation of its ring tone. By comparison, the only view I got of the bird was a few glimpses as it dragged its immense tail through the brush in front of me. Superb indeed!

Despite the sunburn, the instant exposure to Australian bird life in the Royal motivated me to explore further, and the next day I pointed the car inland and drove the couple of hours up to the Blue Mountains. This was a popular weekend spot for Sydneysiders and also a merciful retreat from temperatures soaring close to 40 °C, with a local radio station promising to give away free beer if the magic figure was reached. A couple of days here would also serve a dual purpose; allow me to explore the Capertee road to Glen Davis, apparently a noted birding hotspot in the foothills on the far side of the Blue Mountains, and also help me forget that this was the day Naaz was flying to Switzerland for a Christmas skiing holiday with JJ.

The next morning dawned cool and drizzly and, as a result, my early start on the Capertee road produced some good birds after a slow first hour or so. Once the rain held off and the forest opened out, birds came thick and fast.

There were three more species of Honeyeater, and four more of Parrot, including the local speciality the Red Rumped Parrot, found only in this kind of wooded farmland. I would drive a few kilometres and then stop, birding most of the time from the car, using it as a hide.

In this way I picked up the typical farmland species like Brown Songlark, Masked Lapwing and Peaceful Dove, roadside species such as Willy Wagtail (Australian lack of formality evident here), Hooded Robin, Australian Pipit and Grey Fantail, whilst an unexpected pool yielded ducks like the Hardhead and Wood Duck. My favourite, however, was the Zebra Finch, usually seen in aviaries in the UK, but here all the more resplendent in its pied finery as a flock paraded for my delight on a farmyard fence.

I spent a very happy, bird-full five hours up and down that road, before the mountains loomed again and I turned around for Sydney and home. I was surprised by the speed of the cars using the road, most of it an unmade track. Evidence of the damage this could do was soon apparent; the discovery of two dead kangaroos, one still bleeding and twitching in the road, and the almost tragically comic sight of a wombat upside down in a ditch, its four legs sticking up like the legs of a chair, where the collision with a jeep or farm truck had thrown it. I had spent all morning searching for animals, especially kangaroos, on the road without success and finding ones like this did take some of the gloss off the day.

The weather soon grew hot again, just in time for the Christmas festivities. When I planned the trip I was always nervous of feeling very lonely at this time of year miles from home, and had planned to be in Sydney at that time of year especially so that I could at least be amongst friends. Simon and Cate did me proud, bought me small gifts, made me feel part of the family and generally organised things so that there were always plenty of curious family members around anxious to talk to the limey going around the world on his own. Christmas Day, along with New Year's Day, were the only two days of the entire trip spent on the beach, almost a right of passage for any pale (or by now pink) Brit lucky enough to be Down Under at this time of year. I tried to take it all in, with bikinis taking the place of the James Bond movie as compulsory Christmas Day afternoon viewing, and lunch taken out on the balcony overlooking the sea as the sun began to set. I rang James, my parents and my brother as soon as I thought they would be up, ignoring the expense on the mobile and just happy to hear their voices from so far away. This was the moment I realised just how big the world can be, their Christmas Day beginning as mine ended. It was at this exact moment I thought of documenting all my travels in a book when I got back, my thoughts turned wistful by the setting sun and too much good red wine.

With the kids home from school, a recent death in the family and the impending arrival of another friend from the UK, Simon and Cate certainly had their hands full and so I decided to make myself scarce for a few days, and booked a trip to the Northern Territory and Darwin on impulse.

I had never been to the far north in Australia, and despite it being still the wet season knew that it would yield more good

Osprey overhead: Florida Everglades, USA

Males displaying at the Frigatebird colony: The Galapagos, Ecuador

The endemic Galapagos Dove:
The Galapagos, Ecuador

First sunset on the Equator:
The Galapagos, Ecuador

Me and a 150-year-old tortoise:
The Galapagos, Ecuador

Tawny Frogmouths: a Sydney
garden, Australia

Sperm whale flukeing: off Kaikoura, New Zealand

Tigress and cub: Bandhavgarh, India

birds and also allow me to visit the Kakadu National Park, another long-held ambition.

The long flight north showed me what an immense place Australia is – nearly five hours and still in the same country! However, there the similarity ended; Darwin was damp and muggy, with humidity that stifled you like a wet blanket. The town, a low-rise, low-interest sort of affair, seemed closed and the hotel a lame attempt at a tropical resort. However, I did manage to book myself on a two-day tour of Kakadu starting the following day, arresting any thoughts of being stranded in the city for another day.

Funnily enough, however, one of the best birding spots of this little jaunt turned out to be the park opposite the hotel, although even at walking pace you were sweating profusely unless it was very early or very late in the day. Amidst the aboriginal winos slumped in the rain shelters, I was able to get tropical gems like Rufous Banded Honeyeater, Helmeted Friarbird, Figbird and White Bellied Cuckooshrike, whilst on the adjacent mud-flats there was a massively billed Eastern Curlew and the ubiquitous Lesser Crested Tern. Great birds, but such an effort to find in what felt like an oven turned up to 'High'. It didn't really cool down even after dark.

The air-conditioned bus the next morning was therefore a welcome relief, even if it meant a three-hour drive to get to Kakadu. In the dry season this was good for birds, especially water birds that congregated in the slowly diminishing pools in the area. However, now in 'The Wet', the bird life was much more dispersed and consequently the day's highlights were largely of the non-bird variety: some extremely ancient cave paintings, a flight over the Jim Jim Falls, gigantic termite 'cathedrals', some more flying foxes, a lone dingo on the road and, finally, two wallabies 'boxing' on the lawn of my hotel in front of my room.

The local wetlands did have small numbers of the species they would be inundated with later in the year, especially the trademark Magpie Goose, Radjah Shelducks, and a couple of menacing Jabiru Storks. The hotel grounds turned out to be the best birding spot of the day, however, with a mournful Whistling Kite on the radio antennae and a Barking Owl in a tree outside my room. Bird of the day though must have been the Pheasant Coucal; seconds after the driver informed us of its suicidal tendencies, one decided to ram the bus head-on, making a sound like a grenade going off and leave a huge crack across the windscreen.

The Bird List was only coming along slowly, the summer weather and desire to bird only intermittently in Australia providing a steady trickle of additions rather than a deluge. It was the same the next day – some fantastic scenery, evocating the thousands of years that the local aborigines had revered the rocky escarpments and dense forests as sacred, some great animals, but only limited numbers of birds. If only a fraction of the vast clouds of flies had been birds I'd have loved Kakadu even more than I did.

I eventually warmed a little to Darwin (as if I wasn't warmed enough!), having found the backpacker area and discovered that this area was really lively in the evening, but after a free day swimming in the pool and mooching around in the humidity I was glad to head back towards Sydney and the impending New Year celebrations.

It is a truism I think that in your life you can count the number of really stonking New Year Eve's on the fingers of one hand, unless you are extremely lucky, or perhaps Hugh Hefner. I had dreaded this one, as there is no time like this to reflect drunkenly on life's mistakes. I didn't really need any introspection. I couldn't afford it: had I been right to leave my job? Was my son OK without me? Should I have proposed again to Naazlin? Should I have spent my windfall on a new flat instead? What was I doing here? What was I doing going back?

No point dwelling on all this halfway around the world I thought, and at least whatever we did I would be with friends. Anyhow how bad could Sydney be on the eve of a new year? It was positively great all year in my book.

So on a hill overlooking Sydney Harbour I assembled with Simon and Cate and a bevy of 'rellies' and friends to watch the firework spectacular, and in this setting that was a correct description. Part of me wanted to be partying down at Circular Key with the masses, but it was great drinking champagne around Simon's pool instead, even if it inevitably led to me looking at the stars at midnight and wondering if Naazlin was looking at them too at that moment, thinking of me. All of which conveniently avoided the fact that she was in a different hemisphere and some nine times zones earlier than me. Champagne can do funny things to your brain on New Year's Eve.

My only resolution was that New Year was to be loyal to the original ethic of the trip – to *enjoy* it as a thank you to myself, as something I had given up a lot in order to achieve – and it was almost half over already. I had somehow expected a blinding flash of inspiration by now, prompting a deep and meaningful resolution to return to the UK and do something different with my life, but it never came.

My head was full of my imminent return to London, and my determination to make this count. If there were any chance that Naaz would not go to live in Switzerland, or even just delay it until the summer, than I had to take this opportunity. Also, I planned on inviting her to the final leg of the Trip in South Africa. A two-week holiday with me was going to take a lot of explaining to JJ and, consequently, a lot of persuasion by me.

Over the next few days I worked on Naaz, creating the story that I would be at the Australia v. England Test match at exactly the same time as I would in fact be jetting back to see her in snowbound London. She even told me of her lunch meeting, how it was an exciting opportunity for her career-wise, how strange it was that it was in *this* hotel, one that had a special secret significance to her and I. It was all I could do to not blurt out that I was coming.

I filled in the time trying to get a tan, enjoying a drunken lunch by the harbour with Simon and some old friends, and having a very surprising and impromptu drink with an old girlfriend – a chance meeting in one of the biggest cities on earth, 12,000 miles from home – one of several unnerving coincidences that kept coming up in my time away. She looked less than pleased to see me, as we had split up rather acrimoniously when I started to concentrate on wooing Naaz. I certainly didn't tell her I was about to get on a plane to go back and see her the very next day.

But throughout I had only begun to think of what Naaz would look like when I surprised her in London. Would we still click? Would she still love me? Would she be angry I tricked her? Would the magic have evaporated across the miles that separated us?

These were the questions that burdened me as I sat rather sheepishly in the First Class lounge in Sydney Airport, my bag full of hurriedly bought winter clothes, my head full of carefully crafted doubts.

Those feelings of guilt and doubt that boarded the aircraft with me soon evaporated under a tidal wave of luxury. To travel halfway around the world First Class is the nearest thing I've experienced to positively Roman decadence.

I knew that the actual journey itself was going to be special when, along with my first glass of welcome champagne, I was handed a menu with seven courses of delicious Malaysian food, entitled 'light lunch'. Thank God I'd already sweated off a few pounds in Darwin, because my waistline was about to get a grilling to match that of my credit card.

Somehow I'd always expected my time on the Trip to be spent roughing it in a jungle lodge, travelling on antiquated aircraft with the locals and their livestock. Yet here I was in the First Class cabin, with enough room to play a five-a-side football match with my fellow passengers, the nearest of whom was seated too far away from me to talk to unless I shouted. All a far cry from the usual zoo in Economy Class, I thought.

I spent many happy hours munching on satay, drinking champagne and using my touchscreen monitor to serve me individually chosen movies and audio. Inevitably the question 'Are you making a bloody fool of yourself, Russell' kept reoccurring, but by then I had slipped into luxury mode and recognised that at the very worst this was going to be a great adventure, if an expensive one.

We broke the journey in Kuala Lumpur, and I nearly gave the game away when I rang Naaz in London, pretending to be at the Test match in Sydney, just as the loudspeaker announced that a flight was boarding. Fortunately she could not understand what was said, and I managed to bluff my way through it but it did teach me just how easy it would be to ruin the whole scheme. If it came off, then it would be a story to tell until my dying day, but if I blew it, I would be a laughing stock and, more importantly, a fool to myself.

The second leg was mostly spent asleep, a choice made easier not only by the seat, which seemed to have a hundred and one different positions to aid your in-flight comfort, but also by my new travelling companion, a garrulous and unpleasant British brickie originally from Leeds, and now the worst kind of self-made man living in Perth ('I'm not a snob tha' knows, Russell, but I said to t'airline that they were not picking me up for t'flight in anything less than a bloody Bentley and no mistake.').

Some ten hours later I was peering through the porthole at a dark and wet London town beneath me. The shiny streets, car headlights struggling to penetrate the murk, and miniature Christmas decorations all seemed vaguely familiar, but as if from a time long, long ago. I didn't feel any of the warm glows of familiarity I had expected from my home country, almost as if it blamed me for leaving in the first place.

Heathrow was a chilly and thankless place to come back to, the seemingly omnipresent dark and dank a reminder of how much sunlight and warmth winter steals from us here ever year. I caught the train to Paddington, with the rush hour in full swing in the pitch black, half expecting to bump into somebody I knew. I practised my abbreviated story of why I was back three months early, just in case.

The taxi ride to the hotel in the West End reminded me that London's traffic is worst than most. I still had four hours until the 'meeting' at lunchtime, and managed to glean a couple of hours of fitful sleep so that I wouldn't look too dishevelled and fall asleep in the soup, suffering from jet lag.

The last couple of hours were the worst, excitement mixed with rank terror, heart rate increasing by the minute. I spent inordinately long in the bathroom, trying to decide what to wear from my meagre selection of clothes that had accompanied me, wondering whether I had a decent tan, whether I really had lost any weight.

When the appointed hour arrived I was almost incontinent with apprehension. I rang down to the restaurant to see whether Naaz had arrived; only one of the 'lunch party' had arrived, so everything was going according to plan, but she – that is, Naaz – had decided to wait for her colleague and 'guest' in the foyer.

This momentarily fazed me, as I had always assumed that I would make a grand entrance in the restaurant, and had many times tried to imagine how that first moment would be. I crept down to the lobby of the hotel, hands by now physically shaking, and peered around a pillar, feeling like some sleuth in a detective movie.

Fortunately I saw Naaz straight away, bent low in an armchair and reading up on her notes. Its funny the things that go through your mind at this stage – she has a new jacket, her hair looks great short, her company have played their part and given her fake notes for the bogus meeting – but I couldn't wait any longer and strode across the lobby, to an uncertain fate.

Naaz was so engrossed that she didn't look up as I neared her, and I had an opportunity to plan my point of attack. I decided for some reason to walk up behind her, to maximise the surprise. I stood behind that little figure I knew so well, found I couldn't remember a single planned line that I'd been practising since I left Sydney thirty-six hours before, and simply settled for what I hoped was a deep and intriguing 'Hello, Naazlin'.

You often see in cartoons and comics a graphic illustration of the term 'double take', where the character looks twice to make sure he has seen what he thinks he has seen. Well, Naaz's reaction was the only time I have ever seen a *treble* take in real life.

She looked up, then again and then a third time, as if she could not understand what her eyes were telling her. Meanwhile I was praying quietly that she would be pleased that I was there at all. In an instant she was hugging me close, in the same breath apologising that her boss was due any moment for an appointment with an important client.

It took me a couple of minutes to explain that that *no*, her boss was not coming, and that *no*, there was to be no important client either, she and I were the only lunch guests. It took me until we were sitting at the lunch table to make her understand this, never mind appreciate that I was there at all. We were both shaking like trees in a gale, and then suddenly the tension broke and we embraced in a long, tight and passionate clinch.

The rest of the meal was probably embarrassing for both the

other diners and the waiter, as we were all over each other every time they looked around and neither of us had much appetite other than for each other. She was delighted at my subterfuge, intrigued that her boss was involved, delirious that our chemistry not only seemed intact, but that we were both so clearly exploding with excitement at being reunited after all this time apart – three months that had felt like three years.

I then did go into the one part of my routine that I could remember. I explained that she would be staying the night with me at the hotel, and that after work we could swim in the health club before dinner.

Right on cue, she said she had no swimwear, allowing me to pull out of one bag a purple bikini I had bought her in Sydney. When she then said she had no clothes to go to work with the next day, I pulled from another bag some beautiful lingerie I had bought for her in Darwin. I had all the answers, I was there with her at last and our love was not only intact but also positively throbbing. The entire plan had gone like a dream.

I walked her back to her office, afraid to let her out of my grasp even for a second. Unknown to me, her boss had decided to let the entire company in on the 'surprise', and somehow they had all kept it quiet. The place was in an uproar when I got her back there, and I suffered being the centre of attention whilst she told and retold the story of our meeting to all of her colleagues.

Somehow she explained my appearance to the one person who mattered most, her boyfriend. Naaz has always had this knack of being very persuasive when she wants something badly enough, and a few hours later we were enjoying the comforts of one of London's finest hotels and our first night together after all those sleepless ones apart.

This was only the start of what was to be a fantastic four days together. She cooked for me, we danced in her candlelight, lay awake talking for ever and revisited many of our special places together. The highlight was in Midhurst, and a typically frosty English country-house winter weekend full of four-poster beds, log fires and too much food.

The entire four days went unbelievably well, better than I ever could have imagined, and only once did I think that perhaps I had overstepped the mark a little. On our last morning in Midhurst, with her still dozy in bed in my favourite blue negligee, I presented her with my final gift – a return ticket to Cape Town, to coincide with the last two weeks of my trip there.

For one horrible moment, I thought I had made a terrible mistake. She looked very confused and said, 'Russell, what have you done that for?' Many times I had joked that she must join me, and although I had now given her the means to do just that, I had also taken away her choices; she would find it hard to refuse, but even harder trying to explain it all to her boyfriend.

This was the one part of my scheme that I had not foreseen, and I realised that just giving her a ticket was not necessarily going to get Naaz on a plane to Cape Town in three months' time. I was going to have to make that trip seem the most important thing to her, worth all the heartache and sacrifice as far as JJ was concerned. Somehow I had to keep up the momentum achieved by my sudden reappearance, and capitalise on the depth of our feelings for each other at that point in time. The first thing I did when I got back to Sydney later that month was to ensure that Naaz got a red rose every Monday, counting down the weeks until Cape Town. I hoped against hope that this would help make it as hard as possible for her to bottle out, when the inevitable row with JJ came.

The four days were over all too quickly of course, but at least they did allow us to exorcise one last demon. Our parting back in October had been made all the more painful by the way in which I, afraid to actually say goodbye and unsure how to handle those last moments together, had simply just jumped in a taxi and sped away, with Naaz a forlorn figure in the street out of the back window.

Well, saying goodbye after such an idyllic few days together was never going to be easy, but we both made sure we made the time we needed to do it properly, and to say those things that we hoped would keep our feelings as intense as possible for the coming weeks and months apart. I felt much happier at Heathrow the second time around.

But not entirely guilt free. I agonised over whether to call James again, and wished then that I had found time to go and see him but knew he was too young to understand. Likewise I thought twice before ringing my parents, but did make the call from the First Class lounge to tell them I *had* briefly been back, but was now off again.

Their reaction was pretty typical. Extreme surprise, initial disbelief and then resignation that it was all, after all, for love. My Dad muttered something good-naturedly about having more money than sense; my Mum was suddenly more interested that my return flight would be shared with Rolf Harris. Thank God for understanding parents. The hardest call of all was to Naaz, not finding the words to say how much I had enjoyed being reunited with her and hoping against hope that the next time I would see her would be in her native Africa.

After such a dreamy trip to London, my stopover in Kuala Lumpur was always going to be an anti-climax. I saw it really as an introduction to my time in Southeast Asia in February, and an early opportunity to register some birds on the neglected List from that region (I had agonised over whether to add British birds from the last few days to the List before deciding that would have been a real cheek).

KL (as everyone calls Kuala Lumpur) was an industrious and thriving metropolis, expanding at such a rate in Malaysia's 'tiger economy' that the new airport was built some 20 miles outside the current city, the government realising that in a short time KL would expand itself closer to its new airport.

Along the way, KL had acquired some fantastic roads and impressive modern architecture, but also lost most of its Eastern charm and colonial heritage. Understandable if you were a local, I guess, but like Singapore before it, I felt it was one of those featureless cities only worth going to in order to do either business or shopping.

I spent a day on a tour that took in the last of the British colonial buildings left in the capital, as well as the newer ones, which included the massively impressive Petronas Towers. These dwarfed all the other new towers and skyscrapers in the city and remain the tallest buildings in the world. I felt compelled to do the other tourist stuff, visiting craft workshops and the Hindu cave temples at Batu, but it barely filled my day and my head was full of my time with Naaz.

Meanwhile my belly was full of delicious Malaysian food, one advantage of a country that is the product of so many different cultures. This included an evening in the corner of an authentic Chinese restaurant, where as the only occidental I was stuck on a tiny table almost in the kitchens. It did allow me a glimpse into local Chinese society, however, and I studied with great curiosity the adjoining table, where what appeared to be a local businessman was trying to impress his boss and his wife by buying him an impossibly large feast and getting him to drink around half a bottle of Chivas Regal.

This did at least wet my appetite for parts of Malaysia which weren't like Croydon-in-a-sauna, like Penang, which became another target destination for another time. It also gave me a chance to track down a field guide for Southeast Asian birds. After the best part of a week taking time out from building the List, I could ill afford any more now. I was hungry for Malaysian birds and a chance to play catch-up with a total that was looking more modest by the minute. The problem was that anything representing greenery seemed to have been built upon, at least in my part of the city, and the local parks were largely inaccessible in the little time I had left.

I spent some hours wandering around rather gingerly on some waste ground at the back of my hotel, and managed to get Asian staples like House Crow, House and Palm Swifts, Java Myna, Brown Shrike as well as a couple of less prosaic ones like White Rumped Muna and, especially, a magnificent Blue Throated Bee-Eater. All in all, however, it was an early taste of how in Asia it was much harder to do casual birding, as the cities in particular seemed to have far less species than cities on other continents, like say, Sydney.

And Sydney was back where I ended up, a week after I had left for London. I sensed that my constant coming and going and nocturnal phone calls had worn my welcome a little thinner than I would have liked, and Cate looked almost relieved when I sheepishly announced that I would be spending my last few days in Australia at the end of the month in a hotel in the Rocks, right on the harbour, before leaving for Thailand.

For now, however, I was still the itinerant traveller and I had a few days to kill before leaving (again) for New Zealand at the weekend. Fazed by the whirlwind of travel over the last couple of weeks, I was relieved to spend a couple of days doing domestics and lapping up the sunshine. I wanted to say a big thank you to Simon before I moved on, and it was somehow fitting that my final day as his lodger should be spent with me paying for us both to climb the Sydney Harbour Bridge, the ultimate symbol for the ultimate city in the world.

Ironically, of course, we chose the one day that was not crystal clear and bright sunshine, but at least it made the climb, in full safety gear and in special lightweight uniforms, more bearable. In any weather the view of the harbour from the top was spectacular, with the boats on the water looking like toys from where we were, and the Pacific Ocean just visible beyond the Heads miles away on the seaward side of the harbour. Behind us, you could just make out the new Olympic Stadium at Parramatta, the centre of Sydney's triumphant hosting of the Olympic Games some three months before. It was a fantastic goodbye to a fantastic city.

My final afternoon was spent in Manly, a neighbouring township justly famous for that other Australian icon, the beach. I tried to appreciate a culture built so firmly and simply on sunshine, water and the urge to show off your sexual and sporting prowess, and I do confess that as I relaxed in the sun watching beach volleyball, I very nearly managed it.

The next fortnight was going to be a departure for me. Although New Zealand was right next door, I expected it to be very different to Australia, with its more conservative style and diverse and unique bird life. I certainly hoped so, as my time in Australia, with its mountain of travelling and time spent relaxing over the festive period, had only added just over a hundred species to my total in nearly a month. This would have taken me just a few days in South America, but there really was no substitute for time spent in my favourite city.

But perhaps the biggest difference was the impending arrival of Alison, a former girlfriend and inveterate traveller who would join me for the second week. After so long travelling solo, and so recently returned from Naaz in London, I was already wondering whether different was going to mean good or bad.

BIRD COUNT: 613

8 Parrots, Penguins and Mollymawks: New Zealand, January

I arrived in New Zealand, my fourth country in the space of eight days, on Day 84 of the Trip – exactly halfway through, and with over 60 per cent of my bird target in the bag. Although I was ahead of schedule, my time in Australia (and beyond) had done me no favours as far as the List was concerned, but one look at my new field guide to NZ birds immediately taught me that I would be lucky to make any great inroads into that here.

The twin islands of New Zealand certainly had many distinctive birds, but the very same factors that made these species so distinctive also reduced the numbers available.

New Zealand had never been part of a large continental landmass in the same way as Australia had, and had been isolated in the Tasman Sea for thousands of years. This made migration for birds and animals to the islands much more difficult and, consequently, the List of land birds was unusually small, although this was balanced in part by the multitude of seabird species to be found around the islands.

In such an isolated location, NZ developed an avifauna unique to itself; with very few indigenous mammals, birds on NZ occupied those niches occupied by mammals in other countries, and over thousands of years these species had reacted to the lack of predators by losing their powers of flight. The most famous of these is the four species of Kiwi, predominantly foragers on the forest floor, a niche usually occupied by small rodents.

The problem with this was that man arrived, and introduced these small rodents onto the islands, along with cats, pigs, goats and rats. They cut down the forests and bred millions of sheep on the hillsides to keep the British Empire in wool and lamb. More than any other colony, NZ was developed over the years as a little piece of England (or, more often, Scotland). As a result, by the last century the natural fauna and landscape of NZ had been ravaged on an unprecedented scale.

Unfortunately the introduction of so many alien species had eradicated or extirpated many of the most distinctive species, including some of the flightless parrots and the entire genus of gigantic Moas. Most of the others only survived on lonely outcrops far from the mainland. Instead, the main islands were populated by introduced species from the mother country, and as a consequence this was my first and last opportunity to tick very familiar species from my garden at home, like Blackbird, Dunnock, Greenfinch and Goldfinch.

So much for the natural history lesson. What NZ lacked in numbers of birds it made up for in other natural wonders, not least over a dozen types of Shearwater and Albatross (still called by their Victorian name of 'Mollymawk' for the most part), whales, and raw, magnificent natural beauty. I had arrived from England a week before Alison, and I intended to use that to pick off the prime wildlife sites.

I decided to confine my efforts to the South Island, the more interesting and scenic from a birding point of view. This ensured that with many miles to be driven ahead of me, I would not spread myself too thin. New Zealand was one of the few countries that initially lived up to its stereotype. The drive from Christchurch airport took me past street after street of genteel little stone-clad semis with carefully manicured lawns and well-tended shrubbery. The weather was certainly reminiscent of a spring day at home, and a blessed relief after the cauldron in Sydney.

The impression that somehow I had arrived back in Blighty was further accentuated by the city itself. Not content with being named after the Oxford University college, Christchurch even had a River Avon, complete with straw-boatered punters and strawberries and cream. Compared with the vibrant bustle of Sydney, Christchurch was eerily quiet, its shopping centre full of stores with hardly anyone in them, and everything seemed to shut by mid-evening. The impression was of a well-mannered, English university town on a wet Bank Holiday Monday; a perception not harmed by the busloads of Japanese tourists that seemed to be everywhere.

It was therefore with a little relief that I rented a car the following morning and quickly left Christchurch behind. I had chosen a metallic blue VW Beetle; an idiosyncratic car that seemed ideal for a country that I was already sure was going to give me a multitude of experiences, all of them different. Certainly the car was a novelty outside of the cities, judging by the number of people who stopped and stared as I drove past.

My first stop was the beach town of Kaikoura, nearly a day's leisurely drive north of Christchurch and chiefly famous for its sperm whales. Even avid whale-watchers seldom see this species, as they usually inhabit very deep water. However, here at Kaikoura the continental shelf lies immediately offshore, and every year pods of juvenile male sperm whales hang around the area to feed, and provide the tourist with a unique opportunity to see them in a simple day trip from the harbour. Kaikoura was one of *the* reasons why I came to NZ at all.

That drive down the coast gave me a first insight into the rugged beauty of NZ. The lush farmland and twee towns quickly gave way to a series of barren mountain ranges that divided the island in half, rising to the dramatic Southern Alps further south.

By the time I found Kaikoura, these had grown to an impressive backdrop, surrounding the little beach community and framing it against the impossibly blue Pacific Ocean. Despite being a weekday, mid-January was what passes for high summer on South Island and accommodation was not easy to come by. It took me an hour to secure a bijou little chalet near the beach.

The town centre (as such) was a short drive away, and quickly defined Kaikoura as a funky, individualistic little town, with a small selection of bars and restaurants that invited further inspection. Within an hour I had booked my tours for the next few days, and could afford to relax and begin to enjoy the place.

The rest of the afternoon was spent exploring the beachfront, nearly deserted in a strong wind off the sea, and I hoped that my whale-watching trip the following day would not be affected. I drove out to a point about a mile from my chalet, home to dozens of New Zealand fur seals, all blissfully unaware of the tourists roaming among them taking photos.

I had ticked most of the common bird species from the UK on

the drive down, but here I was able to add really *good* birds like the NZ Pigeon, Pied Shag, White Faced Heron and Variable Oystercatcher. In fact, nearly half of the birds I was to see in NZ were endemic, proof of the country's isolated location.

I had dinner later that night in a great little restaurant, and found that New Zealanders readily talked to anyone, their old-world manners in step with the atmosphere of the country as a whole. Having said that, I spent much of the time talking to Laura, a cheeky young Mancunian who had just started as a waitress there. She was working her way across the country, and desperate for news from home. Before I knew what was happening, and buoyed up by some rather good roast lamb and local red wine, I had surprised myself by asking her out to dinner the next evening, her day off. She surprised me even more by accepting. It was fair to say that I had enjoyed my second day in NZ much more than my first, and it was about to get better still.

Fortunately the wind had dropped by the following morning, and soon after breakfast I had been through the safety briefing, seen the video, bought my souvenirs and had joined over fifty other tourists on the boat out to the continental shelf to find our sperm whales. Whale Watching in Kaikoura was a serious business, it seemed.

Despite the throng, most of who appeared to have just a casual interest in whales, I managed to station myself on a good vantage point on the roof of our boat as we attempted to sneak up on a sperm whale found on our sonar. We slowed to an idle as the whale dived at our arrival, knowing she was most likely to surface again nearby after ten to twenty minutes.

The wait in the heavy swell, feet braced against a ladder so as to be able to use both hands on the binoculars, was a great piece of theatre and designed to set the pulses racing. Just when you assumed the whale had dived deep and moved on, she would appear almost surreptitiously on the other side of the boat, announcing her reappearance with a great sigh of fishy breath. There, she would 'log', just resting on the surface, allowing us to take roll after roll of photos, before taking three or four breaths and diving deep once again.

Several factors made sperm whales particularly obliging for

whale-watchers, like their tall blow and habit of fluking just before diving, but I must confess that one of the most fascinating aspects was how the ancient whalers came to name them.

Having caught their sperm whale, they discovered that the huge bulbous head (think of Moby Dick) was filled with a thick, viscous, white substance – the whale oil that was one of the most valuable products to come from this barbaric trade. However, to the sex-starved whalers on the high seas, the similarity between this and male ejaculate was stark, and so the poor whale was lumbered with the odd name it keeps to this day. Even the fact that female whales also carried heads full of the same substance did nothing to change their minds.

One of the real bonuses of these pelagic trips were the opportunities to see some of NZ's abundant seabirds, and this trip was no exception. It provided the only Australasian Gannet of my stay, plus the local speciality, the NZ White Capped Mollymawk. Logging whales often attract prospecting seabirds, and our two brought in the dainty Cape Pigeon (actually a kind of petrel), plus the extraordinarily ugly Northern Giant Petrel. In addition, our journey back took us past rocky outcrops, which produced Spotted Shags and White Fronted Terns, as well as some dusky dolphins that performed well for us, jumping clean out of the water as they appeared to play alongside our boat.

My date later that evening had put me in a quandary. Knowing that Naaz was still with JJ despite our magical few days together, I was entitled not to feel guilty. I was also under no illusions – Laura was after a good chat and a free dinner, and not rampant sex with a man twice her age. Even so, I was flattered that she'd said yes and excited that, after three months abroad, I'd finally got my first proper date.

In the end my overriding emotion was not guilt or lust, but embarrassment. I thought I had got used to being stared at when I was with Naaz, me being a different age and she a darker colour, but this was nothing compared to the stares I got as we strolled from bar to bar in Kaikoura later that evening. At just twenty, Laura looked like my daughter and, of course, I was old enough for her to be just that. Perhaps it was the small-town mentality, but everywhere we went I imagined the chattering classes to be murmuring 'pervert' under their breath.

Still, a good time was had by all, and Laura was surprisingly good company for a woman nearer to James's age than my own. Good food, great wine, a quick goodnight kiss and it was all over, but it did do my confidence the power of good.

I'd enjoyed yesterday's whale-watch so much that I had immediately booked another for the following day, although the dawn was cold and drizzly and the prospects not nearly as good. The journey out to the shelf was more than a little bumpy, and I spent my time counting those unfortunates who reached for a convenient sick bag before we finally found our first sperm whale. Like a London bus, we then could not avoid them, and found four within close proximity, along with the endemic Fluttering Shearwater and another playful pod of dusky dolphins, who spent fully half an hour in the wake of the boat.

Ever a sucker for punishment, I left one boat to immediately board another, this time a small boat designed specifically to find seabirds in another part of the bay. This they did with the help of an exotic mixture of fish offal and oil called 'chum', which stunk to high heaven, but which to Mollymawks was the equivalent of a double bacon cheeseburger to a starving teenager.

We sailed out leaving a stinky trail in the water behind us, but I was already gutted at missing a Fairy Penguin in the surf as we left the harbour. However, the chum worked its magic swiftly, and within minutes we had attracted a small flock of attendant birds. We stopped a couple of miles off shore, and lowered a small cage containing shark livers into the water behind us, a rich delicacy for all the birds I wanted to see.

They certainly took the bait. The larger Albatrosses – Wandering, Black Browed, and Salvins Mollymawk – were the most aggressive, and took the lion's share of the food, whilst the meeker Royal Albatross, another NZ endemic, stood off waiting for a chance. Meanwhile the sea around the bait cage was whipped into a frenzy by birds jostling for scraps, including Westland and White Chinned Petrels and Hutton's, Buller's and Sooty Shearwaters. As ever around the world, the gulls and terns circled overheard hoping some scraps would be up for grabs. It was a magical experience to be party to this, a human fly-on-the-wall within touching distance of a group of birds that usually spent almost their entire lives out of sight of land.

The entire area had a wealth of wildlife, and it was an easy matter to book a tour to see more of it. The next day saw me on another boat, this time a small rowing boat, on Lake Roratonga. This was an idyllic spot amidst lush farmland, dotted with ancient Maori village sites, where they used to shelter from their cannibalistic cousins who came hunting from North Island.

Here was a great chance to boost the Bird List further, with freshwater species like the locally rare Australasian Crested Grebe and rafts of the exotic Paradise Shelduck, augmented by Australasian Shoveler, NZ Scaup and Grey Duck. Also of note was the Bellbird, one of those rather odd NZ birds with boring plumages but exotic calls. The field guide always listed the Maori names for each bird (in this case, Korimake and Makomako) alongside the English ones. Maori is New Zealnd's second official language, even to the point of being a compulsory subject in schools. And, to sum up NZ birding, I also ticked Song Thrush and Chaffinch on the same day.

I loved the laid-back atmosphere and rich natural wonders of Kaikoura, and it was with a heavy heart that I drove the long journey back to Christchurch. My stay there would not be long, however, for my next trip was to be an equally long one, this time further south to Dunedin and the Otago Peninsula. This area was chiefly famous for its breeding colony of Royal Albatross, the only one in the world of any species not on a small deserted island but, instead, on a peninsula close to one of NZ's major towns.

Driving into Dunedin was a surreal experience, as for all the world you would think that you were in a small Scottish city – Aberdeen, or perhaps Dundee. The town was built on a steep hillside surrounding its thriving harbour. Most of the buildings appeared to be Victorian, and built of granite, and many actually flew the cross of St Andrew, a tribute to the founding fathers. The city centre was as quiet as in Christchurch, despite it being a Saturday, and I wondered if there was some big event elsewhere that someone had forgotten to tell me about. Both the pub I had lunch in and the hotel I checked into had that heavy, soporific, provincial air I remembered from holidays with my grandparents in England in the 1960s, with just a hint of decay about it.

However, there was nothing ordinary about the wildlife on the peninsula. Our first stop was at the centre hosting the breeding Albatross colony. Like most of the wildlife concerns in NZ, this was superbly organised and human interference was kept to a minimum. We waited our turn, and then were ushered into the viewing theatre for our familiarisation lecture, given by a blue-rinsed octogenarian who looked like she would be more at home in the Women's Institute than halfway up a cliff.

Then we were reverently escorted up the remainder of the path to spend our carefully monitored twenty minutes in the large hide overlooking the grassy slopes that were home to the Royal Albatross for that brief period when they stopped circumnavigating Antarctica and came to NZ to breed.

I immediately understood why all the reverence. At close quarters the Albatross was a magnificent bird; although at rest it looked ungainly and awkward, in flight just above the hide, with its 14-foot wingspan, it was utterly majestic. The hide was quiet as we all watched in awe as the birds mastered the wind effortlessly at point-blank range. By way of a bonus, another local rarity, the Stewart Island Shag, also had a small colony underneath the hide's windows.

Stopping quickly to winkle a tiny Fairy Penguin from its burrow down at the beach, and photograph a rather bad tempered Hooker's sea lion disturbed from its slumber, we motored on over to the other side of the peninsula.

About a mile away across the hills and beaches, was a colony of Yellow Eyed Penguins, the world's rarest, and we had to be in position as they emerged from the sea to spend the night in their burrows. The steep climb down was enlivened by the appearance of three duelling juvenile Hooker's sea lions, two of whom were content to play-fight each other in the surf, but the third had discovered that if you chase human tourists they run like hell!

As dusk began to fall, we huddled together in the hide overlooking the Penguins' favourite path from the sea, up the cliff, to the burrows. One by one they emerged from the sea, waddled up the beach, paused to wash off the salt in a freshwater pool, and then climbed with painful dexterity up the cliff face, somehow managing to hang on with feet that were designed for paddling, not climbing. I watched with admiration as one Penguin struggled gamely up the beach far below us, to disappear and then re-emerge, *above* us on the cliff top, silhouetted against the evening sky and strutting doughtily homewards.

Our own exhausting climb upwards back to the mini-bus was made worthwhile, for me at least, by a quick stop on a cliff top to watch long lines of Albatross and Mollymawks – including new species like Buller's – going to roost. It all made for a very quiet sleepy drive home in the dark, but it was worth every minute. New Zealand was proving to be a natural history masterpiece.

I spiced up the drive back to Christchurch by a couple of stops at some gloriously scenic beaches, and picked up a Royal Spoonbill along the way. Not wanting to spend too long in Christchurch, I took a right turn just outside the city and spent the late afternoon on another headland, this time the Banks Peninsula. I was not sure what I was going to find here, except that exploring in NZ was apt to turn up surprises – and this was no exception.

After about an hour of climbing over the peninsula, I suddenly started to descend and all at once the entire tip of the promontory was laid out before me. Driving through little bays and wooded valleys, I reached the charming town of Akoura a little later.

Akoura is unique in that it was the only French settlement on NZ in colonial times, and indeed had kept its Gallic atmosphere in a country that otherwise was more English than England. The cluster of timbered and white-washed houses and the quaint Catholic chapel and wooden lighthouse would have been more at home on the French Riviera on the other side of the world. It was an unexpected discovery and one that I used to full advantage, with a splendid seafood dinner overlooking the harbour.

This brought my first week in NZ to an end, a natural history frenzy and one of the most enjoyable of the trip so far. Now came a departure as Alison was to arrive in Christchurch the following morning and was scheduled to be my travelling companion for most of the second week, before moving north to a friend's wedding in Wellington.

Alison had long been a work colleague, and also a friend for much of that time. She had also briefly been a lover, albeit a casual one when we were both between relationships. If I had not met Naaz when I did there was a good chance we might have taken that further.

Like me, travelling was her passion, and she spent all her spare time and money on it. Super-confident by nature, we shared a sense of humour and a recognition that we were quite selfish in our ways after years of living alone, which at least allowed us to tolerate each other's idiosyncratic ways more than perhaps some others might have done. She was a good organiser and I had jumped at the chance to spend some time abroad with her, in what we both felt was our natural environment.

However, I had never spent that amount of time alone with Ally, so didn't really know what to expect. I thought that we would be compatible, and was certainly looking forward to the company, although perhaps not the fact that the second week would, of necessity, be a compromise that neither of us was particularly good at; Ally was no natural history buff and I was going to have to curb my desire to see birds in order to become a more regular type of tourist for a week.

Ally was one of those women I could talk to, and so she had sat patiently back in the UK whilst I had banged on about this crazy Indian I had met and dispensed well-meaning advice as required. She too had now met someone else, and so it seemed keeping our week together platonic would not be an issue.

Except that life is never that simple. There was certainly still a spark between us, and Naaz didn't help by telling me that she wouldn't be upset if anything were to happen between us, as we were together so far away. The fact that Naaz had started to hang out with an old flame of hers at home, a boyfriend who she had virtually lived with years before and still had a powerful sexual attraction for, confused me further – did she want a licence to misbehave as well, in addition to her relationship with JJ? I certainly wasn't going to condone that, and doubted anyhow whether Ally and I would get into that kind of situation. Suddenly, having another traveller alongside was making life awfully complicated.

However, it didn't seem that way the following morning, when after picking up Ally at the airport, and a quick wash and brush up in my hotel room, our little VW headed for the hills, with its occupants excitedly gabbling and laughing away the miles. It was great to have a kindred spirit on the trip, someone with whom to catch up on all the gossip from home, someone to share all my travellers tales with and a woman who I liked but did not have to suffer the angst of being in love with. She even seemed not to mind when I said I would be on the mobile to Naaz a lot. Mind you, when I said I'd like to stop for birds whenever we could, I could hear her teeth grinding in the passenger seat.

We headed inland from Christchurch, towards the spine of mountains that ran the length of South Island, which included the largest, Mount Cook. Our plan was to see some glaciers, and then head south to Queenstown, with its plethora of outdoor pursuits, before getting Ally back to Christchurch less than a week later. This would involve lots of travelling, but it didn't seem to matter as we ate up the miles in the sunshine.

The countryside was at first rather nondescript but, as we drove on, the far mountains slowly came into view, still snow-capped even in midsummer, and they were to be our backdrop for the next few days. Pretty soon it became apparent that we would be able to see Mount Cook from just about anywhere, its jagged peak dominating its range, peering imperiously down on a succession of lakes along the roadside. Each of them was a curious milky blue, the water turned opaque by the high mineral content. Great for photographs, poor for wildlife, the minerals choking the life out of the lake and few birds in evidence, but impossibly scenic all the same.

By late afternoon we were just outside Twizel, a town less than an hour from Mount Cook, and with Ally beginning to feel the full effects of jet lag, we elected to break our journey here. Twizel had other ideas though – despite looking the proverbial one-horse town, it was full to the gunwales with mountaineering tourists, and we had to pay way over the odds for a room. The day concluded with a good dinner and copious wine, with me almost having to prop Ally up in order for her to stay awake and finish it.

I had wondered what our first night in a shared room would be like, but I shouldn't have worried – Ally wore more clothes to bed that first night than Naaz would wear to go to work. If there were any confusion, our complex choreography to avoid being naked in the bathroom at the same time and the constant late night calls from Naaz left neither of us in any doubt.

We were both glad to leave Twizel behind and soon after breakfast we were at Mount Cook, the scenery and wooden lodges ironically reminiscent of Switzerland. Ally was much happier than I that high winds had cancelled our planned helicopter flight, but we quickly booked a boat trip out to the foot of the nearest glacier instead.

This left me ample time to explore the local bush for birds, if I could take my eyes off the amazing alpine landscape around and above us. This was notable for one of my few successes with 'pishing' – a ridiculous concept where birders make a variety of loud noises, meant to replicate the alarm calls of various species, in an attempt to flush birds out where they can be seen.

This seldom works in Europe, but seems to work all the time in places like the Americas and Australasia, and sure enough a few minutes of frenzied pishing brought a rather startled black-and-white Tomtit out to within a couple of feet. I managed to add a rather drab Grey Warbler and the splendidly named Rifleman, one of the smallest birds of the entire trip, before catching our bus to the meltwater lake at the bottom of the Tasman Glacier.

By this time the weather had begun to deteriorate, with scudding clouds and a brisk wind kicking up dust clouds on the valley floor. It began to look as if the donning of several layers of waterproof gear was a sound investment.

Several orange dinghies took our party out to where the glacier, snaking down from the mountain, had its mouth and was slowly melting into the lake. This was the first surprise; the large amount of loose rock and scree on the mountain had given it a grey crust, and the wall of the glacier rose some 50 feet or so above us like a frozen slag heap, and not the majestic white ice sculpture we had been expecting.

The attendant icebergs, which were originally white or pale blue as they broke off the glacier, soon attracted enough dust to turn them a pale grey as well. As they weathered and melted into strange, rounded and grey lozenge shapes, they looked more like a cough sweet than a glacier.

The most fascinating moment was when we hoisted a large piece of 'Basel ice' into the dinghy. This had broken off the glacier far underwater, and floated to the surface, where it had yet to attract its dusky pallor. We sat there breaking off bits of crystal clear ice, marvelling as we sucked on the fragments that this ice was likely to be at least 500 years old.

The afternoon was spent trying to lose the squalls, and heading south on a long series of interminable switchback roads, slowly climbing into the Southern Alps as we did so. We finally reached Queenstown before I went stir-crazy.

This was where our organisational skills and experience in travelling paid dividends. Inside of an hour we had rented an apartment overlooking the lake, and had booked two tours over the next couple of days. This left the evening free to explore Queenstown, which was a pleasant surprise in a country where generally speaking the towns struggled to live up to the majesty of the countryside.

Queenstown was chiefly notable as the gateway to the Alps and surrounding Fiordland, and the centre for dangerous sports, a typically Antipodean response to the natural wonders around them – after all, this *was* the home of bungee-jumping. Many companies vied with each other to offer the highest jump, off bridges, platforms and even pipelines, along with para-gliding, white-water rafting, river lugeing and all manner of equally crazy and hazardous pastimes.

As a consequence, I expected the town to be like my memories of Cairns in Australia, full of YMCA's and backpackers. I was wrong about that; the town centre was full of posh little bistros clustered around the harbour, the shopping centre had more to offer than simply boots and tents, and it all had an air of refined adventure about it. Certainly some of the best meals of the trip were enjoyed with Ally in Queenstown, the restaurants offering the fusion of European and Asian cuisine that this region does so well.

Our own tastes for adventure were going to be rather more limited, particularly as Ally had to be virtually pushed into the cable car to even go up to the peak to look at the view, one of the tamest of pursuits available locally. It was well worth it though, as one could see right over the town to the far shore and the mountain backdrop beyond.

We then drove alongside the very picturesque lake to the tiny hamlet of Glenorchie, our starting point for a jet-boat ride up the Dart River. All the scenery we had been through had not prepared me for the absolute beauty up here in the Alps. The river was shallow at this time of year, but framed on both sides by wooded hills, which rose to awesome snowy peaks just behind. It was how I imagined Montana to be, but any comparison at all did it no justice – I was not to see a more dramatic landscape in six months away. No wonder the Swiss came all the way down *here* when they needed an alpine backdrop for their chocolate commercials!

It was almost criminal, therefore, to see the Dart River valley the way we did – at very high speed in a jet-boat, complete with handbrake turns which threatened to throw us all in the river, and hair-raising skimming of the sandbanks in inches of water. Great, great fun, and even time for a quick stop to go and see a local trout lake which furnished me with my only NZ Robin of the trip.

I first began to sense friction between Ally and I on the journey back to Queenstown, as we bickered over who was to blame for the fact that we were out of petrol and miles from any filling station. Somehow we got back to civilisation, the car seemingly running on just fumes, but I think the rot had begun to set in.

The mood was not helped by our trip to Milford Sound the next day. This was *the* trip that everyone does in New Zealand, the Sound being one of the most scenic spots in the world, a fiord framed by high cliffs and dozens of waterfalls. Not going here was like going to Venice and not seeing St Marks.

Unfortunately, the term *see* was relative. Milford gets rain on two out of three days in the year, the problem with being surrounded by mountains on all sides, and we must have had more than our ration on the day we went. It rained incessantly all day, and we only glimpsed the splendour of the Sound amidst the wet and the mist, and we had to suffer an eight-hour return coach journey for the privilege. By the time we got back to the apartment, our raw nerves were showing.

But at least the trip to Milford had allowed me to catch up with the fabled Kea, the world's only alpine parrot, and a brute of a one at that, the size of a small dog and with a bill like a Swiss army knife. True to form, it appeared at a stop outside a mountain tunnel, and proceeded to try and rip the wing mirrors off a nearby car, before jumping onto the coach roof and trying to open it like a can of sardines. A great bird, but not one to meet on a dark night.

Back in Queenstown, the fact that Ally and I were good mates came into play. Whilst we had not fallen out, our two strong personalities were rubbing against one another, and our different personal agendas did not help. But because we knew each other so well, at least we could talk about this in an adult way.

I certainly didn't want to turn around and spend two days driving Ally back to Christchurch – we had badly underestimated the distances involved and time was short. I was keen to press on to the West Coast, and travel back across the middle of the island through Arthur's Pass. I had originally wanted to fly further south and try for the Kiwis on the remote Chatham Islands, but had already run out of time for that and anyway that tour was full – a real disappointment.

Recognising our little personal dilemma, and the fact that I was in the middle of a quest rather than a short sightseeing trip, Ally graciously offered to fly back to Christchurch and get a plane to Wellington earlier than planned. The best decision all round, and one that allowed us to have a great last evening together without any of the recent tensions. We kissed goodbye the following morning, glad that we had been able to spend some travelling time together, but also with a little relief that we could now go off and do our own thing once again – in truth, the one thing we both did best.

The following day took me over the ever-impressive Alps to the coast on the other side. Every time I thought I had seen the ultimate vista, the next turn in the road provided me with a better one. Then I turned north, heading for the Wilderness Lodge and Lake Moeraki. I had expected a lodge maybe similar to the one in Ecuador, but this could not have been moredifferent – a well-appointed, luxury establishment with hot showers, gourmet dinners and a well-stocked bar, and populated by the rich US eco-tourists who inhabit such places.

At least it was surrounded by temperate rainforest, and a couple of walks turned up the Weka, another strangely named NZ bird, this one with a collection of startling calls, an iridescent plumage and some diagnostic white wattles on the throat. Other highlights included the family party of Paradise Shelduck on the hotel lawn, the evening glow-worm hunt (which proved that US eco-tourists can make a hell of a racket even in the dark) and a walk out to the nearest beach, with its dramatic waves, lush vegetation and deserted sands. Once again I showed off my newly found pishing skills, bringing down a Brown Creeper for my List.

My final stop before the long drive back across South Island was to Franz Joseph, a small town up the coast at the foot of a glacier of the same name. Here I hoped for another chance to take that flight over the mountain, which the winds on the other side of the range had denied me four days before.

Having found a hotel and booked a helicopter flight for the following day, I spent the afternoon walking up to the glacier mouth, this time a proper white one. From here, you could see right up the body of the glacier to near its source at the top of the mountain. The huge white column had a tunnel at its mouth, where the meltwater trickled out and made a small stream down to the sea. From a distance, it looked like a huge white dragon spewing out lumps of ice.

The neighbouring Fox Glacier was the target for the following day, and I was up much earlier than I needed to be, checking the weather and praying the flight would be OK.

I needn't have worried. By mid-morning we were high amongst the mountain peaks, this time looking down on them instead of up. Up here the snow was pristine, and you could see the glacier on its journey down the mountain. We landed on top of the Fox Glacier and spent an idyllic few minutes looking up into the glare of the snow-capped peaks, and marvelling at how warm it could be at this altitude despite the thick pillows of snow at every turn. Yesterday I had been sweltering on a deserted sandy beach, and today here I was surrounded by millions of tons of snow, just a couple of hours drive and a helicopter ride away. It may have taken four days and a few hundred miles, but I had not only eventually seen a real glacier, I had stood on one!

The drive back to Christchurch could never compete with the splendour of the Alps, and frankly it did not try. A last morning shopping in Christchurch for birthday gifts for Naaz, and I was on the plane all too soon back to Sydney.

New Zealand was a fantastic country, its natural grandeur and unique birds dwarfing the rather disappointing major towns. It was the only place I stayed in on the Trip that I wished I'd had longer to enjoy, to see more of the mountains, to visit the last refuge of the Kiwis, the beauty of Marlborough Sound north of Kaikoura, and perhaps the hot springs at Roratonga and Auckland on North Island. I will definitely go back.

I struggled to reach a bird count of sixty in NZ, due to its rather odd avifauna, but it was one of those places where the quality outweighed the quantity. The Albatrosses and Mollymawks, as well as the sperm whales, provided one of the Trip's real highlights, and nearly half of these sixty birds could only be found in NZ. I can't wait to go back, and this time concentrate on a birding trip that seeks out the rarities in those remote places where humankind has driven them.

My last couple of days in Australia were spent in the centre of Sydney: eating dinner with an old work colleague, my first and last chance to see the harbour lights at night; organising a mountain of Valentine's cards to be sent at intervals to Naaz in London; a final raucous dinner with Simon, after giving a talk about UK media to his company (just about the hardest I'd worked in four months); and all this in one of the most extreme rainstorms I'd ever seen, with over 5 inches in twenty-four hours, ironic after the desert conditions in Australia for most of my stay.

Now came Asia. This would be very different, as I was determined to see as much of the culture of this vast and fascinating region as I could and do some serious travelling, some of it of the backpack variety.

One thing was sure – however different the culture, however delicious the food, however good the birdlife, however alluring the oriental women, whatever adventures lay ahead, it would have to be very good indeed to beat my time Down Under.

And it was.

BIRD COUNT: 673

9 Same – Same, but Different:
   Bangkok and Vietnam, February

Asia was always going to be a challenge. I remembered it fondly from a visit, when I was married, to Thailand, Indonesia and Hong Kong as probably the most different place I had ever been – and it soon became clear as I struggled through a very hot and humid Bangkok airport that little had changed in the fifteen years since. Asia was still the most different place I had ever been.

Bangkok was a good example. Superficially Western, with hundreds of miles of freeways, skyscrapers and shopping malls, it had nonetheless retained its peculiarly Thai culture despite this.

To walk the streets of Sukhumvit, especially after dark, was to drink from a vast well of mysterious and exotic potions. The humidity was ever present, but seemed to seep into every molecule after dark. The dim street lighting gave flashing glimpses of street life – laughing prostitutes, mutilated beggars, whole families living right there on the street, and the claustrophobic traffic seemingly always nose to tail, with its soundtrack of blaring horns, Thai pop music and bellowed oaths.

The smells in Bangkok were almost as evocative as the sights and sounds – sate grilling on a tiny kerbside barbecue, the smell of sweat and disinfectant from the brash Girlie bars that were everywhere, the sweet jasmine from the roadside florists. Even today I'm not sure whether I loathed or loved Bangkok, but I sure as hell will never forget it.

Life was certainly lived out there on the streets as it had been for centuries, despite the grandeur of the air-conditioned palaces they call hotels in these parts. Inside these five-star monoliths you could be anywhere, but outside you could only be in Bangkok.

There were other, less poetic reasons why Asia would present a challenge. First, I knew that I would have to work hard for my birds in Asia, and had not really scheduled any time specifically for birding as yet. This despite the fact that my tick rate in the last six weeks Down Under was about half that of that in the Americas beforehand. I was only just ahead of schedule now, and I had to add some new species to that List, and soon.

However, during the next two days touring Bangkok, I failed to see even one new species. I steeled myself for some hard birding in between the tours I had planned.

Secondly, the tempting city of Bangkok was to be my home this year on Valentine's day, and also on both Naaz's birthday and my own. I privately wondered exactly how I would deal with the distance between us at such an emotional time of year, and unlike Christmas, dates that were far more personal to us both.

However, my first three days in Bangkok were simply an appetiser, as I was soon to embark on a ten-day tour of Vietnam. This, and a later tour in Thailand to the northern city of Chang Mai, were my only attempts to truly rough it on the entire Trip. Both tours were of the backpacking variety, where I would be rooming with others in the party in cheap hostels and local houses, eating local food, using local transport and meeting local people. All very different to the *bourgeois* version of travelling I had grown accustomed to, and a sample I think of how my journey would have been had I done it in my early twenties rather then my early forties. I was already looking forward to it.

But for now, I had three days to quickly revisit all the must-see sites in Bangkok that I remembered from years before, and I lost no time hitting the tourist trail. I started at the phenomenal Royal Palace, a complete confection of Siamese architecture, with its apparently random collection of spires, dragons, gardens, Buddha's and huge distinctively gabled palaces, all clad garishly in white stucco with millions of pieces of coloured glass and gold paint to reflect the ever-present sun. It was as if Hansel and Gretel's gingerbread house had been transported to Asia. God knows, I thought as I looked around in awe, what opium can do to an architect's mind. Over the top it may have been, but you couldn't help but be blown away by the sheer, concentrated intensity of this ornate extravaganza. The afternoon was only spoiled by the sheer volume of tourists, and by the carefully planned 'unscheduled' stop at a gem factory on the way home.

The Thai's sheer audacity when it came to commercialism was one of the things I remembered well from my first visit, but it didn't stop me still being surprised as I walked around Sukhumvit after dinner.

This area was not only one of the big centres for hotels in Bangkok but, partly as a result of that fact, was also one of the centres for the sex trade in the city. The Thais had a way of de-stigmatising the entire thing, as if their very innocence and earnest industry made it seem simply like any other local business.

You could sit at any bar along the street and see why the trade was based here. There were legions of men, most of them white and middle aged, with their arms around elegant, slim, beautiful Thai girls, many of whom were less than half their age. This never-ending jamboree was a source of nightly amusement, and I concluded that although some were obviously the result of financial transactions, others had to be the product of the many marriage bureaus, catering for Westerners, that seemed to be everywhere.

Some, though, seemed much more innocent, with a lonely Western guy looking for romance and some fun, and a poor teenage Thai girl, probably having moved to a Bangkok slum from the countryside, looking for a 'wealthy' Western boyfriend to look after her and her family.

Whatever the motive, the one overriding feature of these relationships was how bored both parties looked. Outside the bedroom, they had little in common, could not converse easily (the girls seldom spoke much English) and perversely reminded me of those bored middle-aged couples you see at home, eating together in a restaurant and never exchanging a word. So much for the magic and adventure of the Orient.

My one visit to a local club – well, you can only body swerve around the delectable girls (and ladyboys) outside the club *so* many times before curiosity gets the better of you – proved my point.

Inside, groups of very bored, very young Thai girls pretended to dance to the music, for all the world looking like those polar bears in the zoo that sway on the spot, out of their minds with boredom. Occasionally one of the equally bored middle-aged Westerners would buy one a drink, and disappear into the night with them – in a different context, perhaps he might look like a father taking his young daughter home from a party, but here, well, business is business. It sounds awful and tawdry now, but back then it just seemed normal. Was this rather sad business any worse than leaving these girls to rot in poverty in the rice fields? Who are we to say anyway?

I must say I was tempted, especially as I had begun to sense that, back in the UK, Naaz was now being hotly pursued by that ex-boyfriend of hers and I had begun to doubt her powers to resist his muscle-bound charms. Later that month I was scheduled to spend more time in Bangkok, and wondered whether I would continue to be as well behaved.

The next day was spent in information overload, as I was bussed from temple to temple and saturated in dates, anecdotes and dainty morsels of Buddhist folklore. Impressive they certainly were, but it was amazing how quickly you stopped being amazed by yet another temple containing yet another kind of Buddha. A lazy afternoon in my hotel's rooftop pool was just what the doctor ordered before my long journey to a very different kind of Asia the following day.

And Hanoi certainly was different to Bangkok, in just about every way imaginable. To start with, the airport was completely deserted, and contrasted deeply with the organised chaos in Bangkok where thousands of people were moving through on their way to other parts of Asia. Here, nobody seemed to be going anywhere.

My taxi into the city was virtually the only car on the road, but this did not stop my driver from constantly blaring his horn at every rickshaw and bicycle that came within range. Here in the countryside, almost everything was moved by bicycle – entire families, pigs for market, bundles of upside-down chickens, king-size beds, I saw it all. The most comical sight of all was of a family of four on a motorcycle: two small kids at the back holding on for grim death; a mother putting on her make-up looking into the wing mirror; and, right at the front, a small child of eight or nine wearing shades and gripping the handlebars as he drove the rest of his family into town.

Meanwhile the rice paddies alongside the road showed Asia as it once was, with industrious peasants labouring with what looked like ancient machinery or lumbering water buffalo, and conspicuous by their huge conical hats. Vietnam was a country full of wild contrasts already.

Once in the city, the traffic jams were virtually all of bicycles, or motorbikes with a minimum of two passengers. Out walking the streets, the experience became more surreal by the moment. The whole place seemed to be painted as if by a dirty brush, with all the colour and vibrancy of Asia somehow diluted by a wash of grim austerity, and not helped by the equally grim autumnal weather this far north.

As a Westerner, I soon found that wherever you went in Vietnam, those people that weren't openly gawping at you as if you were a circus sideshow, were desperately trying to sell you something, and the further north you were, the more desperate they seemed.

A typical encounter would be for a young teenage boy or girl to wander up to you and start a conversation in broken English. If it was a girl they started by asking if you were married, if a boy it would start with what a good player Michael Owen was. Inevitably this soon turned into a request for help to buy school books, then the sale of some tatty T-shirts and, on more than one occasion, a shrill 'F*** your mother!' when I firmly refused to buy anything.

However, it was mostly good-natured, and as I took a rickshaw around town I began to see that Hanoi had a certain dilapidated charm. The one thing that Vietnam did have in abundance were war museums, as befits any nation that had seen off the Japanese, the French and the Americans during thirty years of constant warfare.

The Army museum in Hanoi was pretty typical, with lots of horrific photographs, decaying maps of famous battles and a commentary which was unashamedly propagandist, usually talking of the 'courageous revolutionary forces of Father Ho destroying the Imperialist aggressors and their quisling forces' (the South Vietnamese). In fact, pretty much every city we were to go through had a rusting heap of American tanks or aeroplanes outside a museum of some kind.

Our hotel was probably quite handsome in its day, with big, old teak doors and a grand staircase. Even now it stood out a little as it had hot water that worked (some of the time) and television in the rooms, both regarded as luxuries in the north of Vietnam it seems.

That evening I met with my fellow travellers on the ten-day whistle-stop tour of Vietnam. Our guide was a red-headed Aussie/ Canadian, with a rather funky and irreverent air as befits a tour composed mainly of travellers under the age of forty – I was one of the oldest on the trip. I soon realised that in my four months abroad I had been totally cocooned from the *true* traveller and their culture, which was never more apparent than in the backpacking havens of Asia.

Many of the party had already come through the cowboy country of Cambodia (fantastic but dangerous, apparently), and the backwoods of Laos (unspoilt for now, and never to be pronounced with the 's' on the end, I learnt). They had already learnt the ropes for fighting off beggars, haggling in markets and finding charming restaurants in out-of-the-way places, and it left me feeling a little unprepared. They also had a whole series of catchphrases, the most famous of which was 'same – same but different'; apparently everything was SSBD, and I soon found myself using it all the time without ever finding out its true meaning.

My room-mate for the tour was Andrew, a beefy Aussie of Greek extraction, with lots to say for himself (most of it rather amusing), but still living down a scandal in Laos where two local girls were found in his room. These tours were all about not exploiting the local cultures and not exporting our own to them, and so this little piece of local industry was seen as an anathema. Two Norwegians had been sent home in disgrace, but Andrew somehow clung on in there. Our time together was never going to be less than adventurous, I thought.

After our first dinner together, I endured a rather tedious evening at the famous but overrated Water Puppet show, considerably enlivened by a rickshaw race back to the hotel, at heart stopping speeds in the heavy traffic and dim street lighting. I was already learning to adapt to being with these new friends every waking minute for the next ten days.

Our first stop was to Halong Bay, a UNESCO world heritage site further north near the Chinese border, and some four hours through the suicidal traffic on poor roads. The resort primarily catered for Chinese tourists and, consequently, was rather primitive by Western standards, although being expanded in every conceivable direction it seemed as old travel restrictions (and national enmities) disappeared. However, the town was dwarfed by the grandeur of the surrounding bay.

The huge area was littered with small limestone islands, each topped with lush green forest, and many eaten away by erosion into a succession of strange shapes. There was something vaguely prehistoric about the place, and its image was further enhanced by a flotilla of tiny houseboats, each with a string of washing and an endless supply of charming toddlers aboard (all begging for money, naturally). Only in Asia could you so easily come across a slice of life like this at every turn and continue to be amazed. What a pity that they thought nothing of desecrating their uniquely beautiful surroundings, throwing all their rubbish overboard for it to float past us, mocking our wonderment at the place.

Our boat chugged across the wonderfully photogenic seascape, and finally gave me my first Asian birds, this my fifth day on the continent. Many of the islands seemed to have resident Black Capped Kingfishers and Blue Rock Thrushes, a species I was used to seeing thousands of miles away in the Mediterranean. A stop to look inside a limestone cave was made more interesting by the Japanese White Eye and Stripe Throated Bulbul in the surrounding scrub. However, the commonest bird were the dozens of Black Kites, to be seen everywhere picking up morsels from the surface of the sea from the rubbish thrown overboard.

That evening we had a rather boozy Vietnamese dinner, with the local liquor being passed around the table, as the motley collection of Brits, Aussies and Kiwis got to know each other. My final memory of a great evening was singing (I use the term loosely here) 'Like a Virgin' with two of the girls and a bunch of Thai tourists at a local karaoke bar on the edge of the bay. So this was what backpacking was all about!

The journey back to Hanoi was not so much fun, driving rain ruining the chances of birding from the mini-bus. However, the greatest impediment was not the weather but the fact that every inch of the lowlands in Vietnam were given over to agriculture, and no plot was too small to have rice growing on it. The exceptions to this were the graveyards, considered sacred to the Vietnamese, and so each rice paddie would typically have a small hill in the middle, covered in gravestones and with a small temple on top.

I still expected more from the watery landscape than just two types of Kingfisher – Pied and Common – and I suspected that the millions of tons of bombs and chemicals dropped on the north in recent times had taken their toll of local wildlife, not to mention the local custom of shooting everything on sight for the pot – more than once we saw a gaggle of men with rifles hungrily pursuing something in the paddies. My Vietnamese List was not going to be a big one, though I badly needed it to be.

A free afternoon in Hanoi was a chance to explore the Old Quarter, hidden away in the grimy labyrinthine streets near the main lake in the centre of the city. Here I browsed a local market, and once again pondered upon Asian attitudes to the living creature around them, as I saw chickens having their throats cut, crabs boiled alive and frogs having their legs removed, all in the name of cuisine. It was churlish, I felt, to abhor this, as tonight I would have been quite prepared to sit down and enjoy a meal made of those very same animals – cruel, but necessary, I guess.

Hanoi was ridiculously cheap by our standards, and this included the few Internet connections. This was just as well, because along with being the cheapest in the world, they were probably the slowest in the world as well. I got back in touch with Naaz after a few days' absence, during which she had gone away for a supposed innocent weekend with her ex. My guts churned at this news, even though I was not even her boyfriend. The prospect of sharing her with two men, if it came to that, was a like a severe headache that would not go away. I tried hard to forget about it, as I could do nothing to change her mind, but it was never far from the surface most of the time while I was in Vietnam. It was hard to believe that only a month before we were so happy together, and now this all seemed to be falling apart.

We began our long journey south that night, taking the Reunification Express train to the old Imperial city of Hue. Thank God we got the more modern of the two trains, especially as I had celebrated being in close confinement with my fellow travellers (as I had aboard the *Cachalote*) by coming down with a heavy cold. This was presumably the result of swapping 40 °C in Bangkok for 15 in Hanoi, and I tried to keep it at bay with my small stock of medicines and umpteen tangerines.

The journey was made all the more dramatic by the sudden disappearance of Andrew. Amidst all the rumours, it was clear that he was a more brooding character than his garrulous exterior would suggest and carried a big grudge from the fracas in Laos. In addition, he was under enormous pressure from his Greek family to get married (he was thirty-one), and had taken the extraordinary step of leaving a tour he had already paid for, in order to find a Vietnamese wife! Stranger than fiction, but at least it guaranteed me the only single room for the remainder of the trip.

When we awoke from fitful sleep, the landscape had begun to change. The rice paddies were still there, but the formerly ubiquitous water buffalo was now a rarity, replaced further south by more modern machines. There seemed to be something rather more pastoral, a little less desperate here. I even managed to pick up Javan Pond Heron, Grey Wagtail, Burmese Shrike and Yellow Wattled Lapwing from the train – as many new birds in the first half an hour as I had managed in the previous five days!

Although at first seeming like a huge shanty town; Hue was obviously of some economic importance, but it had lost much of its Imperial architecture when Ho Chi Minh had forcibly deposed the Emperor at the end of the Second World War– the communists were nothing if not thorough. He then had the perfect excuse to move the capital to his home town, Hanoi. We had little time to dwell on this, however, as we sped off on motorcycles, each clinging grimly to our local driver upfront.

We at first visited two temples, but unlike those in Bangkok these were of the small village variety, and the second of them cooked a superb lunch in our honour (same – same but different, indeed). We spent the afternoon bombing around the countryside (but not in the same sense as the Americans had thirty years before us) and driving through various little villages where the people were dirt poor but seemed unbelievably cheerful. Everywhere we went we were accompanied by a gaggle of smiling children, all running and repeating the one word of English they knew – 'Hello,

Hello, Hello . . .' Some of them didn't even ask us for any money.

In one village a man stopped our motorcycle and insisted I try his homebrew, which from his rather dishevelled appearance I concluded he had been sampling himself on a regular basis. It tasted like hell, but I was not drunk enough to accept his second offer, which was to chew on something he held in a bowl and which looked vaguely like it may have been attached to an unidentified animal at some point long long ago.

That afternoon was one of the highlights of the entire time in Asia and, it struck me, an experience I would only have had on a tour such as this. The chance to see the locals at work and play, making bricks, planting rice, playing with their kids, was something I will never forget. How noble and content these people were, despite living four to a room in some dishevelled shack and having to supplement their meagre farming income by making conical peasant hats for tourists. It gave you an insight into what true wealth really is.

This was reinforced at dinner that night, when we all sat down to a plate of delicious noodles cooked and served by the chef and his gorgeous daughters, in a small room with just a couple of electric lights which passed for a local restaurant. He had been profoundly deaf and dumb since birth, but was probably the happiest man I've ever met, and so very proud of the wedding pictures of his eldest daughter. As we all played football with a shuttlecock in the street afterwards (including all the girls) I began to realise that Vietnam's appeal lay in the beauty of it's people rather than of it's countryside.

The next day was totally free, and I elected to take it easy as I had been frenetically travelling for most of the last ten days. The weather had cleared along with my cold, and I visited the local citadel, or those parts that the communists had left standing. Even now there were soldiers everywhere who did not want you in certain parts of the palace. Only a few buildings were intact unfortunately, and twentieth-century politics had robbed us all of what was once a palace to rival that in Bangkok. The most interesting find, however, was a very welcome Daurian Redstart on the city walls, probably a wintering bird from China.

I had an interesting encounter that lunchtime whilst eating

noodles once again with our deaf friend and his family. A small boy sidled up to me and passed me a note. On it, in three languages, was a request for me to look at his coins, all salvaged from the grounds of the palace opposite.

Feeling sorry for the little chap, I duly pored through his collection, with James back at home in mind; I felt he would be especially knocked out if I bought some coins for his collection from a boy who looked about the same age as he. Many were rather grubby, and he explained that they could be cleaned in Coca-Cola (I found out later this was true), and that he hadn't already done so because tourists would think they were all fake.

I have to say I admired the kid's business acumen, and bought a handful from him. I was still amazed, though, when he pulled a sheaf of banknotes from his pocket in order to give me change.

'What's a little kid like you doing with all that money?' I asked.

By way of explanation as he pocketed my cash, he passed me his identity card and pointed to his birth date. I read incredulously – 15 September 1968. I had been fooled by a thirty-three-year-old man, closer in age to me than to James! He correctly assumed, I guess, that I would not have bought from an adult, and so made the most of his stunted growth, squeaky voice and child-like face. Vietnam just got stranger and stranger.

I went to Da Bang market that afternoon, looking to buy a sleeping sheet for my tour in Thailand. True to form, by the time I had haggled over the price of a silk sleeping sheet, I had been seduced by the charm of the owner's two young daughters, both diminutive in stature but big in the customer relations department.

They took me on a tour of the market, and before I knew what had happened I was in a roadside shack, having a body and scalp massage, manicure and pedicure. Against my expectations, I was not offered anything else, except the usual request to 'help with their school books'. It was with considerable reluctance I had to say no, feeling slightly aggrieved that they would try this after all the trade I had given them. That's business in Vietnam, I thought.

These little dramas followed us south in Vietnam. The long drive south on the laughingly titled Highway One – surely one of the worst roads in Asia – was interspersed with a breakdown with a bad fuel leak, a contretemps with a gaggle of local traders and a bad case of diarrhoea amongst the group, that needed a doctor's intervention. Eventually we arrived in Hoi An in the early afternoon, with the weather warming up as soon as we crossed the Hai Van Pass into the south.

Hoi An was always the favourite stop on the tour apparently, a charming compact little town that looked exactly like a Westerner's idea of what a small provincial Vietnamese town *should* look like – lots of small houses, many a cross between Chinese and French architecture, and clustered around a picturesque harbour on the riverfront. It was here we had another humorous introduction to The Vietnamese Way. As the party sauntered along the road alongside the river, the local boat owners began shouting out to us in a variety of languages, an approach that we all had got used to in Vietnam.

One of our crew responded to one shout at him in English and said rather politely 'Not today thank you'. Quick as a flash came the only other word of English the guy must have known – 'Bollocks' he said with a huge grin. And I'm willing to bet he knew that word in half a dozen languages.

Hoi An was known not only for its old-world colonial charm, but also for its tailoring, with a greater density of seamstresses per square metre than anywhere on earth. Most tourists get something made here and none of us were exceptions. However, it was one of those 'well it fitted in the shop in Vietnam' moments for some of my clothes and, although I made good use of some of the others, they have remained at the back of my wardrobe ever since my return, never likely to be worn again.

There actually appeared to be some nightlife worthy of the name in Hoi An, although the local communist council insisted that it all closed down at 11 p.m. Hence I found myself enjoying another sumptuous dinner (my waistline was beginning to regret this by now), having a few drinks with some Americans from another tour group in a local bar and sauntering home alone after closing time in the unlit streets, again the object of some interest from every passing local.

Its funny how just a few days changes your attitudes on this kind of trip – I would never have considered this kind of thing a week before in Thailand, but after sampling so much of the local life in Vietnam, it now seemed the natural thing to do.

The next morning I promised myself a boat trip up the river, in a desperate attempt to find some bloody birds. I now know of course that all the decent birds in Vietnam are in the mountains, but at this point this was still a mystery and I spent a completely barren two hours on the water without even a hint of a new bird. However, this trip did offer up another rather large slice of real life.

Men might have sold trips on the sampans on the river, but it was women who did all the hard work, here and in the rice paddies. My driver in the stern was a women who looked old, although in practice should could have been anywhere from forty to seventy, except for one thing; sitting on the prow, spasmodically paddling when she felt like it, was *her* mother, with more lines on her face than the London Telephone Exchange and a grand total of three teeth. This made about ten teeth between the daughter and her combined, and all stained a lovely shade of deep purple from chewing beetle nut all day, as was the local custom. To walk the streets of a Vietnamese town was to avoid the little piles of purple spittle that dotted the pavements everywhere.

As we made our sedate way downstream, passing fisherman and houseboats, the old woman behind me grumbled away in her singsong language, whilst ahead of me the *really* old woman grumbled also; as far as I could tell they were not speaking to each other, rather having their own private grumble out loud. I was grateful when suddenly the Ancient One stopped paddling and promptly fell asleep on the deck.

This being Vietnam, we were always going to make an 'unscheduled' stop at some place selling something, and sure enough halfway through the trip we pulled in to the bank next to a tiny homestead with a small kiln attached. To be polite I thought I would at least have a cursory look, but all thoughts of such niceties disappeared as it became obvious that the steep bank here was made of wet mud and as treacherous as hell. Sure enough, I ended up sliding down said bank into the river up to my knees, and would have felt even more aggrieved had I not been tickled by the sight of the two aged crones rolling on their backs, kicking their legs in the air in hysterics. I didn't buy anything, though, on bloody principle.

Our last night in Hoi An was taken up by a cooking lesson, and we carefully made our own spring rolls and steamed sea bass with noodles before wolfing the lot down afterwards. Vietnam was turning into something of a culinary festival.

Day 111 saw me pass the two-thirds completed point for the Trip, and with it several realisations. First, I had barely thought of home for ages, save for some of the people I was missing. Secondly, I was no longer ahead of schedule on the List – was I going to fail after all? Thirdly, it was of no comfort to learn that the rule that applies to your two weeks in the Med. also applies to a six-month travel odyssey – namely, that once you get past halfway, time begins to flash by at an alarming rate.

Our last leg of the tour bypassed what remained of Highway One, and we took the notoriously unreliable national carrier to Saigon – few people even today call it Ho Chi Minh City.

This was the one place which reminded me of Thailand, in that Saigon was on its way to becoming another sweaty zoo of a metropolis. At least here they had retained some of the best of the French colonial architecture, including the post office, railway station and a cathedral rather bizarrely called Notre Dame – every brick had originally been transported from France. I noticed with a wry smile that the Madonna had a blue flashing neon halo and pondered on how religion manifests itself around the world – you couldn't imagine *that* catching on in Paris.

Saigon's War Museum was a masterpiece in displaying how cruel a war can be, and what carnage the most powerful country in the world can inflict on a rural peasant nation. With typically Asian disregard for decorum, it included glass cases containing the remains of stillborn babies caused by the mother ingesting Agent Orange, and close-ups of the horrific burns caused by napalm. It was impossible to watch this in anything other than a hushed silence, and as a Westerner, with a small pang of guilt that, with Americans and Australians involved in the war, were we all somehow responsible for this. Who could blame the Vietnamese for revelling in their unlikely victory? Their story of overcoming and surviving such suffering is an example to the world.

Later in the week I would visit the Cuchi Tunnels, a tunnel

network built right under a US Army base where the Vietcong operated for a number of years, resisting all attempts to evict them. Whole families would be born, live and die there, and you only had to be underground for a few minutes, or to eat their staple diet of cassava, to marvel at the fortitude it must have taken to survive down there.

Ever the tourist, I paid my five bucks and used up the magazine of a US M16 rifle on a specially erected target. As with any man, the kick of the rifle and the loud report appealed to something primeval in me, maybe the same hunter-gatherer instinct that led me to search for birds all day every day. But I felt so guilty that I would do this for fun, when in the not so distant past (for example, our driver's father had been crippled in the war and imprisoned after it) this would have been a matter of life and death.

Still this was nothing compared with Cambodia apparently. One of the group told a story of how an American tourist had bought a rocket launcher and single rocket in a local market, and then bought himself a live cow. Then he launched said rocket at the poor cow, missing in the process. Life was incredibly cheap in this part of the world.

Saigon has a staggering two million motorcycles, and it seemed as though we met every one of them as our coach struggled to leave the rush hour behind the next morning en route to the Mekong Delta.

In insufferable heat, we took canoe rides, visited a local bee farm and a coconut candy factory, held gigantic snakes for the cameras and had our first putrid meal of the tour whilst listening to a local folk group. It was all rather touristy and, of course, although I had expected to finally add lots of extra birds to the List, I could manage only three – Pied Fantail, Chinese Pond Heron, and a fine pair of Olive Backed Sunbirds. There were others deep in the bush, but they annoyingly flitted out of range whenever the binoculars got anywhere near my eyes, it seemed.

Back in Saigon, after a particularly vigorous massage by a pretty Chinese girl, it was time for the group's final meal together. Our leader had picked this vast tent which served barbecue's of all shapes and sizes, a pleasant change of diet I thought as I wolfed down the *steak-frites*. At the front of the tent were cages full of pigeons, sparrows and eels. So *this* is where all the bird life is, I thought. The cages held at least one species I had not seen and probably more, but despite how short of birds I had become I still could not justify ticking these poor creatures.

I bought everyone a cocktail at our last stop, a posh bar on the roof of a five-star hotel, our first and last taste of luxury on the tour, and with a great view over the city. Our group had gelled well, the abrupt departure of Andrew notwithstanding, and I had grown particularly fond of Sophie, a young Australian, although too young even for me. We had enjoyed a strong bond whilst travelling, were both carrying some emotional baggage, and I liked her strong individualistic streak and independent spirit – she was about to leave by herself to travel in the mountains alone, despite rumours of armed rebels taking Westerners hostage. Not for the first time, I found myself wishing I were fifteen years younger.

Undeterred by a typical bag-snatching incident – petty crime was a big growth industry in Saigon – we parted amidst tears and (unfulfilled) promises to keep in touch. Such is the way with so many people you meet travelling, ships that pass in the night.

I stayed on an extra day, visiting Cuchi, and then the backpacker area back in Saigon. This was an opportunity for a marathon email session, and also for testing my ability to resist the bug which had afflicted the group all week with a very welcome curry, accompanied unsurprisingly by one of Saigon's regular power cuts. Pestered relentlessly by cyclo drivers offering promises of 'boom-boom' girls, I was finally glad to leave Saigon behind, although not so Vietnam.

I was fascinated by the entire country of Vietnam, of the contrast between North and South, city and country, old and new. Still badly scarred by the war, this was a country that had been through a lot and was only now beginning to rise from the ashes.

Although the food was a real highlight, the people made the trip, and compensated for the fact that on our tour we could not say it was a beautiful country (although I have heard the mountains are fantastic). I had certainly had enough little adventures to fill my journal from Vietnam, and had a real laugh with my travelling companions. It left me wondering whether I should have done more of this kind of travelling elsewhere.

The only disappointment was obviously the almost complete lack of bird life. I managed a paltry fourteen species in Vietnam in ten days – a truly pathetic total. This small number was, of course, compromised by the huge amount of travelling and by the nature of the tour, but it certainly gave me renewed vigour to get some new birds back in Thailand.

As I flew back to Bangkok, and looked down at the ravages of Agent Orange I could still see as grey scars on the lush green of the Mekong Delta, I played Vietnam back in my head, and enjoyed once again all those funny moments with the locals. Thailand would be very different, but probably how Vietnam would be in twenty years' time, Cambodia in thirty and Laos in forty. I was privileged to see it now.

Meanwhile I had the spectre of Valentine's Day and a birthday amidst the fleshpots of Sukhamvit to consider, not to mention some serious birding to catch up on. How was I going to handle that?

BIRD COUNT: 687

10 A Thai around my Neck: Thailand, February

The traffic from the airport to my hotel was even worse than I remembered it, except unlike Hanoi or Saigon there was not a motorbike to be seen – it was mostly big gas-guzzling limousines, a status symbol in a consumer-mad society. And, whereas in Vietnam you could somehow cross the road by closing both eyes and just going for it – this is the first thing that you are told as a tourist and unbelievably the motorbikes *do* actually miss you – here you were condemned to sitting in long queues of cars.

This did give me a chance to dwell on the huge billboards, skyscrapers and high-tech offices along the freeway, and compare them to the resolute low-rise, low-tech cityscape of Hanoi, or even Saigon for the most part. And, I still didn't see any birds apart from some rather forlorn local pigeons. It was a constant mystery to me, how I spent over a week in the Thailand's capital city and yet did not log one new bird there.

My first job back in the air-conditioned comfort of my hotel was to go into Valentine's Day overdrive. This was my one big chance to really impress Naaz and let her know how much I missed her today of all days, and how much I wanted her to be on that plane to Cape Town in just over a month's time.

The usual frenzy of emails and phone calls to London as the sun set magnificently behind me was broken only by the arrival of a card from Naaz delivered to my room, and of a huge bunch of orchids from me at Naaz's office – a real feat this to find these in winter in England and, consequently, the most expensive flowers I will ever buy. Still, at least they outshone JJ's boring old red roses.

It was thus a very loved-up and buoyant middle-aged tourist who went out for dinner that night in Sukhumvit, and this mood protected me from the despair of seeing so many Europeans having a young Thai girl to spend their Valentine's night with.

I figured an early night would be a safe course of action, but it was as I returned through the lobby that I saw the girl. She sat there in a raincoat, looking terribly lost and lonely and trying to read a magazine. Still full of self-confidence, I swallowed hard and started a conversation with her. Without being pretty, she had that alluring yet innocent quality that oriental women seem to exude and, what's more, seemed to have been let down by her friend who she was due to meet.

Feeling sorry for her, I invited her for a drink in the local bar, and at once found all those curious eyes boring into my back as I sat there with a girl half my age. All was going swimmingly until after about ten minutes she explained she had nowhere to sleep, and that if I were to help with a little money for her mother's operation she would sleep with me.

My first reaction was 'Well at least that makes a difference from money for school books', but my second was to realise how tempted I might have been had it not been for those loving calls to Naaz earlier; that, and 'Ahh, so that's how it works'. Either way, I felt no guilt as I finished my drink and left her sitting in the bar alone. Another one would be along for her in a minute.

The next day was a busy one – I had to prepare for my tour to the north, and organise the spare time I had left in Thailand, hopefully with birds in mind.

By lunchtime I had booked a coach tour to the Bridge on the River Kwai the next day – a right of passage for anyone interested in history but a journey I had missed on my last trip years before. It also offered the prospect of getting out into the countryside and, hopefully, getting some long overdue Thai birds – my total for the country was currently zero.

Someone on my travels had mentioned that Krabi, down near Phuket in the south, was a good place to see wintering Siberian waders and, armed only with that snippet of information, I decided on the spur of the moment to spend a few days down there later in the month. It involved a couple of flights and a stay in an expensive hotel, but I was getting desperate and the spectre of returning home without 1000 birds was beginning to loom large.

More than any other country I had visited since I left South

America, I knew that Thailand could be great for birds, that is, if I had come to the country to *just* see birds. The remaining rainforest held innumerable rare and unusual Pittas, Tailorbirds and Drongos, and the hill country some exciting wintering species from further north, but I had arrived as a regular tourist and now turning this into a birding trip whilst in Thailand was proving impossible.

So, instead of tracking some rare Pheasant in the jungle, the next day saw me on a coach to Kanchanaburi, and the River Kwai. It was a sorry testament to the passage of time that I felt none of the anger and disbelief that I experienced in the war museums in Vietnam at the cemetery for our own dead at Kanchanaburi. Even a ride on the Death Railway and a chance to see the appalling deprivations inflicted by the Japanese on allied troops failed to move me as much as I expected, terrible though it all was. I imagine that the Vietnamese war was more real for me, as I could remember it from my childhood, and the Second World War seemed like ancient history by comparison. The one statistic that stayed with me from the Bridge was that many more local peasants died from forced labour and starvation than any allied troops did. You were never told that at home.

But, I did at least finally break my duck (if that is the right phrase here) and registered some Thai birds at long last – a flock of Red Throated Barbets atop the trees alongside the Kwai river, a dashing Oriental Turtle Dove overhead, and both Long Tailed Shrike and Green Bee-Eater on the wires from the train. And about time too.

None of this prepared me for what was to be a rather unusual evening. It all started so tamely too, with a good dinner with two of the people on the Vietnam trip, both now staying in my hotel. When they retired to bed, I went to a local bar, usually rather less predatory than the others, and had a quiet beer whilst I took in the local colour, of which there was never any shortage here.

All the bars had a number of bar girls, whose only role seemed to be to keep the Western customers happy enough to keep buying drinks. Usually I was happy to be by myself rather than be pestered by some false bonhomie, but somehow Key was different.

She came and sat next to me, and rather than ask my age or ask if I was married, said she wanted to practise her English and would I mind if she chatted to me.

During the course of the next hour or so I became intrigued by Key (like all Thais, she had shortened her real name, which would often run to a half a dozen words or more). She was twenty-three, studying English at college and had left behind an unhappy home – this too seemed typical, as many of the girls I had chatted to had parents that had broken up, usually because the man could not keep it in his trousers. Despite being devout Buddhists, your average Thai man (and woman) did not equate faith with fidelity in the way we would in Europe.

Key was believable, and did not once ask me for anything never mind offer it. She was not especially pretty, having a round face without the high cheekbones of most of the girls, but was incredibly cute, and I could not fail to notice that her slim, taut body was offset rather nicely by a belly-button ring and a large tattoo of a snake on her flat stomach. I don't know how it happened exactly but, before I knew it, I had not only told her that it was my birthday in a couple of days but I had arranged to meet her to celebrate it the following evening!

I awoke excited by the prospect of a *real* date with a Thai girl. With JJ in London to celebrate Naaz's birthday (the day before mine) I also told myself I didn't have to feel guilty about it. I spent the afternoon on the Chaopraya River, negotiating with a local boatman to take me downriver, around the *klongs* or canals, and then stop at Wat Po, an imposing tower of a temple unlike any other and set majestically overlooking Bangkok's murky waterways.

As we swept gently through the canals, I managed to tick White Vented Mynah, Oriental Magpie Robin and Spotted Dove, but already my thoughts were on a much more exotic bird later that night.

Key arrived bang on time, contrary to my expectations as I half expected her not to appear at all. She looked ravishing, wearing a black off-the-shoulder number that was cropped short so that once again the snake on her belly was impossible to miss.

I had let her choose where we went for dinner, and rather fetchingly she had chosen a particularly cheesy river cruise, mostly full of drunken Koreans. The food was average and the wine worse, but the Chaopraya was far more impressive at night, and Wat Po looked magical when lit up. We got on well, and I found myself flirting madly with her and holding her hands before the dessert arrived.

I don't know what I expected from the encounter, except a fun time to help celebrate my impending forty-third birthday, which would actually be spent on a train to the north in a couple of days. Like Laura in New Zealand, I knew that Key was here mostly to have a good time on my money, but I didn't mind. It was ever the same the world over- and not expensive anyway.

We went from the boat to a Latino club, but the fear of making an ass of myself on the dance floor, coupled with the knowledge that Naaz would be with JJ in exactly this type of club in London in a few hours time, drove us into a basement disco in a local hotel instead.

This was much more my type of place, with music I sort of knew how to dance to and full of yuppie locals and Westerners with Thai girlfriends . . . just like mine, I told myself with a smug grin.

We drank lots of cocktails, smoked two enormous cigars and danced and danced. Key had become very tactile and we embraced and even kissed briefly, but the one thing that bothered her (I later found out) was not the racial or age differences, not the cocktails or cigars, or even the kissing. It was the fact that I loved to play with her hair, and to touch a Thai's head in public is regarded as impolite.

We seemed to be heading inexorably back to my hotel by the early hours, and I cannot say I minded very much. It was no surprise, however, when she said she would meet me in my room in a few minutes as she didn't want to be thought a hooker as we entered the hotel together. 'Well' I thought to myself 'that was great fun, but that's the last I'll see of her' and sure enough I was still waiting alone in my hotel room fifteen minutes later.

I decided that this was an exciting enough opportunity to at least look for her, to see whether she had fallen foul of the hotel security, but suddenly there she was, having bought a tiny birthday cake and four cans of beer as a surprise!

Now, modesty prevents any further details in a book such as this, but lets just say that I was to see her snake at much closer range later that night . . . and that we didn't need a plate for that cake.

I was to see Key for lunch once later on that trip, and we exchanged phone numbers and email addresses (as you do). The fact that she emailed me twice before asking for cash to help her rent a flat, did nothing to ruin a memorable evening, and merely reinforced my belief that sex and money are inextricably, but not necessarily cynically, linked together in Thailand.

So it was with a particularly self-satisfied air that afternoon, plus a hangover and a suddenly vulnerable digestive system (perhaps we should have got a plate for that cake after all), that I got a cab through the afternoon traffic to meet up with my next backpacker tour, this one starting in the Koi San Road, heart of the Bangkok backpacker district.

Here was a very different Bangkok to the one I'd known. Full of Westerners, but this time dressed almost exclusively in football strips from home, or looking stupid in bandanas and sarongs. Most were trawling the local stalls for tattoos, hair braids or more exotic substances no doubt. I suddenly felt older than I ever did the previous evening, and it was no sacrifice at all to retire early to bed with grumbling diarrhoea and leave all the garish and brash delights of Koi San behind me for the night.

It was a strangely fractured mutual celebratory call to Naaz the following morning – Key was still on my mind, JJ had just left her bed and I had got nothing except a dangerous bottom for my birthday. She had been expecting me to misbehave in some way in Bangkok, and the news that I had (I spared her the full details) both pissed her off and vindicated her weekend with her boyfriend all at once. I was glad to join our first team meeting downstairs before it all got a bit too complicated.

My companions for the second of my two backpacker journeys were not greatly different to the first – perhaps a bit younger and funkier, with a couple of Canadians thrown in to spice up the Brit/Aussie/Kiwi rivalry. Not much of a surprise that neither tour included any Americans, since they were thin on the ground throughout my travels except on their own continent.

This seven-day excursion would be another whistle-stop affair, taking in the old Imperial capital of Sukhothai and the northern towns of Lampung and Chang Mai. Once again we would be simply scratching the surface, but I hoped we would see more of the 'real' Thailand in the way we had with Vietnam.

It was certainly a strange birthday, spending almost all that first day on the train as it ambled through the never-ending Bangkok suburbs and eventually out to the rice paddies beyond. It was a great chance to get to know others in the group and to find out whether first impressions were misconceptions or not, but by late afternoon dehydration, a rising temperature and boredom had rendered me speechless. Still, it gave me plenty of time to look for birds as we sped quicker now through marshy terrain – in this way familiar birds like Greenshank and Black Winged Stilt, as well as unfamiliar ones like Red Wattled Lapwing, Eastern Marsh Harrier and, best of all, some Asian Openbills, succeeded in keeping me awake and my bowels at bay.

We arrived at the nearest station to Sukhothai, still an hour's drive away in the back of a songthau, the typical truck-cum-bus that is the habitual transport in rural Thailand. The cool evening breeze blowing through the open sides was a blessed relief after the claustrophobic train, and we amused ourselves waving to the female cyclists, all clad demurely in long dresses and hats on their way back from college, as the sun set over the steamy landscape.

We reached the guest house that was to be our first stop after dark, and were allocated a bamboo hut between two. I roomed with Jeff, a rotund Canadian serviceman who had been to every continent on earth and tried to find hamburgers in all of them. When he told me he didn't like rice I knew he was in for a rough ride. Our rooms were basic to say the least, with water in a big bucket to shower with and mosquito nets trying to offer some protection against the myriad of bugs assembling after dark. I wouldn't have minded except for my rather delicate state, but my mood soon brightened when a birthday cake appeared, especially made for me by the owner – a lovely touch so far from home, and a yummy accompaniment to our substantial Thai feast.

The next day promised to be a test of my constitution, as we were to cycle a couple of miles to the ruined temples of the old capital. By the morning, however, I felt a little better and enjoyed our guided tour of the temples, long since looted of all their finery by the Khmers, but still impressive and amongst the oldest in the country – some 700 years old compared twith less than 200 for the Royal Place in Bangkok.

The giant Walking Buddha still had an elegance that came down across the centuries, although some of the magic was inevitably lost amidst our guide's strangled command of the English language. What a shame we could never have seen it at its pomp, with all the jewels and precious stones and metals decorating each statue – a sight to see for sure.

As it got hotter and hotter, so the temples lost some of their appeal and we had a welcome break for lunch in a traditional family teak house, built on stilts to provide shade underneath and decorated, like all the local houses, with dozens of family photographs and pictures of the Thai Royal Family. By the time we had fulfilled our obligations to the local batik and silver makers, we were all ready for that cold shower and a late dinner, after which I could log today's birds – Hoopoe (a much richer cinnamon than the European equivalent), Indian Roller, Brown Prinia and, at the guest house, a Rufous Tailed Tailorbird. In these temperatures birding was going to be hard, especially as we were constantly travelling, but at least we were heading in the right direction.

Jeff was turning out to be a hoot despite his eccentricities as a 'Newfie' (from Newfoundland). In particular he had a passionate hatred of all things American, (excepting hamburgers of course) and a keen desire to travel everywhere, as long as it involved maximum comfort and minimal effort. He made me look like Ranulph Fiennes. In general though, the group was slow to gel, with more singles with their own agendas than in Vietnam and a few too many spiky egos. I guess when you go on a backpacker tour you take pot luck with your companions.

The morn saw us once again on the move, and with yet another change of transport. We crammed onto the bus for Lampang, with the overspill passengers occupying tiny plastic seats they had put in the aisles. The guy in the aisle to my right must have been some 25 stone. It was a surreal experience watching a Mr Bean video on the coach (the silent humour translates well for the Thais, who seem to love him) whilst zipping through the Thai countryside, and actually *feeling* the guy next to you sweating.

Our accommodation in Lampang was equally as rustic as the

last, with us all bedding on the floor two to a room again in a very old and very long teak house. At least the showers were hot, and my digestive system was showing signs of life.

The benefit of these days spent largely travelling was that it allowed us to cram plenty of activity in the following day, and by lunchtime we had already watched a rather naff elephant show at the local conservation centre, but then had the privilege of taking a ride on these marvellous creatures. It seemed faintly demeaning to have reduced these intelligent beasts to the role of tourist ride, but as many of them were rescued from a life of hell in logging camps up in the hills, I doubt that they complained, especially when compared with those that had lost legs or feet in landmine accidents along the border with Burma.

I was grateful that I had not taken this ride earlier in the week, as a combination of the swaying motion and our animal's incredible flatulence would have been too much to bear.

The afternoon was spent with a Karen hill tribe village, watching them weave and marvelling at how clean and well ordered it all looked compared with the Thais themselves, especially surprising given that they were a nomadic tribe who had to be coerced into staying anywhere. The women wove bright garments of cloth on handlooms that reminded me a little of the Andean Indians in Ecuador, a million years before (or so it seemed). The children in particular were beautiful, and I was a little sad that the tour's ethical policy apparently prevented me from giving them even one sweet − a lack of any dental care whatsoever meant that this was regarded as dangerous.

Our final journey of the day was to the local hot springs, which had been developed as a resort and packaged together in a curious mixture of Disneyland meets a municipal swimming pool. Certainly the springs were hot (you could boil eggs in them) and, if you could stand the smell of sulphur, refreshing.

By now the group had got used to seeing me with a pair of binoculars perpetually in my right hand, and this had proved especially fruitful in the farmland around the Karen village, picking up Pied Bushchat, Oriental Skylark and Paddyfield Pipit, but it was still a trickle and what I needed was a flood.

The never-ending journey saw us bedded down that night in a

private house just outside of Chang Mai. Here we all slept on the floor in one communal room, the nearest I'd come to slumming it on the entire trip, especially given the primeval state of the toilets and showers, but the magnificent feast afterwards and hilarious lessons in Thai dancing from the family's children and school friends made it worthwhile.

This was the night in which I also discovered that you don't have to be middle aged to snore in bed, nor in fact was it the strict preserve of the men on the trip. The girls may look angelic in daylight, but after dark they can make a racket with the best of us. It was finally a relief when the Mosque next door began calling prayers on a giant megaphone and shook us all out of bed early the next morning. It certainly helped me spot some Thai garden birds before breakfast – Ochraceous Bulbul, Chestnut Flanked White Eye and the marvellously gaudy Scarlet Backed Flowerpecker.

An early start once again saw us sweating on bicycles, touring the local farms and countryside and, at one memorable stop, teaching English to a classroom of Thai children – a magical experience. By the time we left they all seemed to be experts in hangman (our version of teaching them English) and the words to the entire collection of Spice Girl songs.

The short ride in our songthau took us into Chang Mai that afternoon, and gave us a chance to indulge ourselves in Thailand's second city. For the girls this meant maximum shopping opportunity, for Jeff a chance to try the local McDonald's, and for me my usual email and phone frenzy, having been parted from Naazlin for nearly a week – one which seemed like a year after all the emotional trauma of the last weekend in Bangkok. Exactly a month until I hoped to get her on the plane for South Africa, and out of the clutches of not only JJ but also all her other admirers, if only for a couple of weeks.

Our final dinner together as a group never quite took off, mostly because we were all too exhausted by a week of relentless travel. The group still had not really bonded, wheras in Vietnam we had gelled quickly the night we all got smashed at the karaoke club in Halong Bay. If ever I was to organise such a tour I think a drunken get-together on the first night would be compulsory.

We were to leave the following afternoon by overnight train to

Bangkok, and I took full advantage of our free morning by walking around the old city walls, taking a ride in a rickshaw and visiting some impressive temples, one of which sold me a miniature bronze Buddha evidently hacked from a pagoda at some point in the past. Just when you think you are 'templed-out', you turn a corner and there is an even more impressive one.

I also had the misfortune to see a street vendor pester some tourists into buying a small basket of sparrow-like birds called Scaly Breasted Munias, in order to set them free. Somebody had already told me that these vendors put a tiny amount of heroin in the bird's drinking water, thus ensuring that they will always come back to their owner, to then be packaged up and sold again.

Since I had bought a bigger rucksack in Bangkok, I had changed my mind about souvenirs, especially as travelling in Asia without buying mementoes was both unthinkable, and given the Thai's flair for commerce, impossible. A flurry of T-shirts followed therefore that afternoon. Part of me was sorry that, like many of the foreign tourists in Chang Mai, we were not to head further north to trek in the hills, but I was also looking forward to simply *not* travelling anywhere for a bit. But first I had to brave the boredom of the fourteen-hour train journey home to Bangkok.

It was almost dark as we left Chang Mai, putting paid to any hopes I had of repeating my birding experience of the journey north earlier in the week. We each had bought some kind of food for the journey – sandwiches, crisps, fruit, plenty of water. However, the regular travellers on this route were certainly not going to settle for such basic fare, and the Thai family opposite me mercurially pulled from a small bag an immense fish, which covered the entire table between us. It was slowly demolished by hand, with much smacking of lips and picking of teeth with the bones, and in the end I was grateful when the attendants came round and pulled down the beds from their overhanging lockers.

The swaying of the train, the squealing of the brakes and the constant hubbub from the party of Japanese schoolgirls next door meant that it was a broken night despite our exhaustion, and we crawled into the Bangkok rush hour the next morning drained, both physically and metaphorically. Our farewell breakfast, and then those curiously emotional goodbyes – with people you've

175

only known for a week and will never see again – and I just had time to change and repack before my flight down to Krabi and, hopefully, some serious bird watching at last.

I arrived at the new airport in Krabi feeling that I just wanted to vegetate by the pool for a few days, especially having seen my hotel – a palatial resort on the river and surrounded by the distinctive elliptical limestone outcrops that pepper the area.

I really didn't need this amount of luxury at this stage and would have settled the other side of town at the small beach resort where all the backpackers stayed. But now I was here for my final few days in Thailand, I decided I might as well make the most of it – even if I just about had the place to myself. Or perhaps that should have been *ourselves* since, once again, I was sharing the hotel with a couple from the tour I had just left, another bizarre coincidence. They made good dinner companions on the first night, anyhow.

I was raring to go next morning, however, for the first day dedicated to birding in a month or more. No wonder that bird total is described in my journal at this point as 'paltry'. I had no idea where the main harbour was, but I took a taxi from the hotel hoping to rent a boat and a captain for the day, and get them to find wherever the area was where all the good shorebirds hung out.

This was easier said than done. Krabi was not yet a tourist base, although it was expanding rapidly with the help of its new and larger airport runway. It looked and felt like a provincial Thai town, where few people spoke English. Plus, it wasn't even on the sea, being a couple of miles upriver from the nearest mudflats.

I finally found the help of a friendly local tour operator, who negotiated with the local boat-owners on my behalf and explained that I wanted to see birds, not beaches. I was soon seated in one of the traditional Thai longboats, with a long prow ahead and a propeller on a long pole behind. My guy spoke very little English, and was so dishevelled and toothless that I doubted he could last a day in the sunshine, but I was told that he knew where the birds were, and that was more important.

We headed upriver, away from the sea, at first since the tide on the mudflats was too low and we therefore had a couple of hours to kill. By the time I had all this arranged, the day was already very hot, and there was little to see beyond the admittedly wonderful scenery, with rocky gorges competing with limestone pinnacles for my attention.

I did manage to spot a pair of Striated Swallows hawking insects over the river, and a small flock of dumpy Dusky Crag Martins high up over one of the small mountains, plus the ubiquitous Blue Eared Kingfishers and Common Sandpipers that were easy to find. But I was itching for the tide to turn so that we could get amongst the more interesting species out on the mudflats.

We finally made it out that far around lunchtime, and at first I thought we had blown it once again as I couldn't see any mudflats at all, anywhere. However, as we sailed further out, I began to see the rows of bamboo fences that the fisherman had erected offshore, and sure enough as we got closer still I could see that just about every pole had a bird on top of it.

Here were more birds than I had collectively seen all month in Asia. I got my dentally challenged captain to sail up and down the fences, so that I could pick them all off – the most common being the Lesser Crested Terns, with their large orange bills, but it was the clusters of waders that I was most interested in. At low speed, and with me frantically flicking through the field guide for winter plumages of Siberian waders, I managed to tick off common species familiar from home like Common and Spotted Redshanks, and Eurasian Curlew. Then, as I got my eye in, the really valuable stuff – first a couple of Great Knot and then the major target species, Nordmann's Greenshank. They stood out from their more numerous cousins, the Common Greenshank, by their much paler plumage, although an individual already coming into breeding plumage, with its boldly mottled chest and spangled back, sealed identification for me.

This was enough to keep me going for the journey up the absurdly picturesque coast, and right through our noodle lunch on Buddha Island, especially with the Brahminy Kites circling overhead. This island was a little bit of paradise, with fine white sand, water warmer than bathwater and dramatic rocky outcrops offshore, only spoiled by the determination of some of the resident

Germans to bathe naked regardless of how much more than a 100 kilos they weighed.

The tide was ebbing again by the time we got back to the bamboo poles, but I persuaded my captain to hang around long enough so that I could sift through the birds on the now extensive mudflats. Here were some more gems, especially my first Terek Sandpiper, a species I kept missing in Europe. You could add to this Lesser Sand Plover, another elusive find at home. I was so excited by this long overdue glut of birds that I forgot to feel guilty when, as we tried to leave, the boat ran aground on the sandbanks, and even as my skinny little captain was pushing me and the boat towards deeper water, I could not take my eyes off the flocks of birds all around.

'Welcome back!' I smiled to myself.

Birding around the hotel was another pleasant surprise, as nowhere else in Asia had I found a place where you could just casually try to find local birds and succeed. The grounds were especially lush close to the river and before dark I had added Ashy Drongo and its sartorially elegant cousin, the Greater Racket-Tailed Drongo, along with Greater Coucal and Ashy Minivet. Early morning was better still, and before I left I had added over another dozen species, such as Plaintive Cuckoo, an iridescent pair of Glossy Starlings, a tiny Grey Capped Pygmy Woodpecker and the formidable Chestnut Bellied Malkoha.

I had run out of time to do a proper trek into the rainforest in search of Pittas and the like and, not wanting to go to sea again, I thought a bit of active adventure might come in handy for my last day in Krabi. I was definitely not fit enough to trek, and too scared to white-water raft, and so elected to spend the day kayaking around the local sea caves.

This involved a major road trip along the coast, picking up our 'league of nations' of fellow kayakers along the way, and changing songthaus when the first broke down. Once in the water, kayaking was easier than it looked, especially sharing a boat with my partner for the afternoon, a Belgian aircraft designer. A roll of the shoulders, dig in with the paddle blade, push . . . and we were off.

Sandwiched either side of yet another stupendous lunch, this one complete with fresh coconut milk, we spent a hot but happy afternoon paddling into caves resplendent with stalagmites and stalactites, in some cases so large that we were paddling in the dark before emerging into a sunlit grotto on the far side. We got out to explore one on foot, not the easiest exercise in sodden shorts and wet sandals, but then again its 3000-year-old cave paintings were older than most of those in Kakadu, if not as impressive.

My final morning involved what, for me, was a very rare activity – a swim and a read by the pool. It felt great to kick back and not do anything at all after all the frenzied travelling in Asia, but I was not sorry to leave for the airport – the hotel was a lonely place and my head was already full of thoughts of India, just around the corner.

One final night in the sinful capital of Thailand and I was packing all my scattered belongings, souvenirs and dirty laundry for the long flight to Mumbai and India – still Asia, but likely to be very different, I thought. I was not wrong.

My month in Thailand and Vietnam had been short on birds but long on adventure. It felt good to have seen both countries as a traveller rather than as a tourist, and my encounter with Key reinforced my view that I had actually *lived* in Thailand, rather than simply passed through.

That late spurt in Krabi had finally taken my Thailand bird total to just over fifty, but it was still not good enough for the List, although almost all those birds were life ticks for me.

My decision to tour in Asia rather than bird, as much as the absence of bird life in general, had put further strain on the total, and coupled with my wanderings in Australia and New Zealand it was now looking sorry for itself; it had taken six weeks to get to bird 250, eight weeks to get to bird 500, but *eighteen* weeks to even get within sight of bird 750. It had been a long, slow struggle, and with only six weeks left I was behind schedule for the first time on the Trip.

I was relying heavily on the fact that most of my time in India would be devoted to wildlife. India would come up trumps for me – it simply had to.

BIRD COUNT: 742

11 Planes, Cranes and Automobiles: Corbett and Bharatpur, India, March

From the point at which I had queued in the chaos of the Indian Embassy for my visa back in the UK months before, I thought I knew what to expect from India – noise, crowds, colour and bureaucracy. And from having a relationship with an Indian girl, I thought I knew what to expect from the people – passion, energy, directness and complexity. But nothing prepared me for this conundrum of a country once I actually landed in Mumbai (or Bombay as I'd thought it was still called).

That first night in the country was a decent guide to my experiences in nearly a month there. My driver and guide picked me up at the airport, and we launched into the maelstrom of humanity that seemed to occupy every square inch of space. In the gloom of the shanty town surrounding the airport, all I could make out were thousands of Indians in shorts and little else; jammed onto buses, three to a bicycle, sleeping on the pavement, squatting in the gutter, eating with their fingers.

One aspect of this first impression was not to change throughout my stay – the vast majority of the people on the street, driving my cars or acting as my guide, working in shops or in the hotels, were men. The only time I saw more than the occasional woman or two was when passing through a rural village, where it seemed women did all the work. This was in stark contrast to where I'd just come from – the public face of Thailand, for example, was always female orientated. Those of you reading this carefully will know which I preferred.

We crept along at a snail's pace and then suddenly, amidst the squalor, there was an oasis of calm and light – my hotel. We parked outside the impressive frontage, and already I was glad to be away from the dark mysteries of Indian street life. Then came another salutary lesson in travelling in India – although we had come precisely half a mile from the airport terminal, the guide asked me to 'say thank you' to my driver. I was going to say a rather cheery 'Ta, mate' when his outstretched palm left me in no doubt as to what was expected.

This was an unwelcome aspect of being a tourist in India. Everywhere I went I felt I was being hassled for money, not in a charming way as in Vietnam, but with a sort of surly belligerence, almost as if it was an unpleasant smell left over from colonial times. It was not an auspicious start.

The hotel itself was as familiar as the outside world was strange to me. There was MTV on the television, obsequious waiters and overpriced food in the restaurant, and a huge commission to change any money. There, from my first night in India, was the clash of East meets West that was to make my stay so very memorable, but also so very unnerving.

Daylight saw me make the journey back to the airport and catch the shuttle to Delhi, amazed at just how many people seemed to be on the move in this vast country. I was certainly unprepared, on a domestic flight, to have two security checks inside the airport, and yet another on the tarmac itself, as if to underline that with so many people comes the friction of race, religion and, in India at least, class.

I had been warned as to the unreliability of airlines in India, but that first journey was uneventful despite the obvious age of the plane. My new guide in Delhi briefed me on my itinerary, which was long and complex, involving lots of road journeys and transfers between the various nature reserves ahead. I had already resigned myself to the fact that something was bound to go wrong at some point.

Like many Indians I met, my guide was more interested in life in England than in showing me India, and was studying hard in order to move abroad. In the cities at least, people seemed busy trying to shake off their Indian identity, and the young Indians I encountered could have passed for Americans of the MTV generation at a stroke. Only the faded grandeur of the hotel spoke of the Englishness of the Raj that I had expected.

I decided not to waste my half day in India's capital and hastily organised a city tour. I had yet another guide for this and I was already thinking that my supply of rupees was insufficient for all the tips I was going to have to cover.

He duly turned up, resplendent in his white suit and curiously dyed hair that gave it an alarming purple tint in the bright sunlight. He rattled off his patter in a deep resonating voice as if he was reciting Shakespeare on stage, complete with multiple rolls of the tongue and grand gestures. He didn't take kindly to me interrupting his performance with my usual barrage of questions, but I felt if I was going to have to reward each and every one of these guys with a tip then at least I should dictate the agenda.

We first went to Gandhi's Tomb, an understated onyx plinth in keeping with the modesty of the man. Around it lay garlands of chrysanthemums in the holy Hindu colours of yellow and white, whilst an eternal flame burned nearby. My guide told me the story of 'The Father of the Nation', cruelly murdered not by the British or by the Muslims, but by a fanatic from his own religion. Only in India could religious tolerance wear so thin.

The visit to the Red Fort was obligatory, but also welcome, and here was ample evidence of the time the British spent in the country. Equally telling, the scrum outside the fort was oppressive, with hundreds of street vendors almost demanding you buy from them, umpteen snake charmers and beggars thrusting tiny, filthy babies at you with a constant refrain of 'ten rupees mister, ten rupees', or pointing to their mouths and saying 'chapati, chapati'. Not for the first time I felt massively guilty as I turned my back on them. You pay one, I reasoned, you have to pay them all.

My guide's behaviour during this onslaught was another lesson learnt. Rather than offer advice, or shepherd me to safety, he bought a Coke and stood back and watched, as if to say 'hey we all need to make a living mate – you're on your own'.

Next stop was a Muslim mosque nearby, one of the most important in the country. I had expected that inside there would been oasis of calm, but once again all human life was here. My guide asked me if I would like to see the relics of the Prophet that were kept in the mosque, and rarely shown to visitors. My immediate thought was how great it would be to tell Naaz that I seen such holy Muslim relics, especially the 'The Beard of the Prophet'.

He managed to find the keeper of these antiquities, and with great drama the tiny room housing the relics was opened up for me. Instantly I was nearly trampled in the rush as dozens of schoolboys (again, all males in this mosque) crowded around, buffeting me to the back until my guide very sternly and solemnly told them where to go.

The relics were reverently presented to me – a weathered leather sandal, a footprint set in *stone* (and of a different size to the sandal) and, most implausibly of all, some hennaed strands of hair set in glass. I was, of course, invited to say a special thank you but had not understood the full importance of these relics; for what I thought was a fair tip to the Imam who had presented these treasures was looked at with some disgust and instantly thrown over his shoulder. Nice man, I thought.

Afterwards I wondered, if I had believed those relics were real, would I have offered more money. I suppose if they were authenticated in some way, or at least been more believable, the answer is most definitely yes. Certainly if they had been Christian relics I might have taken it more seriously, but these were produced in such a matter of fact way, plus I was already so used to being hustled in India, that my immediate conclusion was that it was another scam.

Perhaps I had just been shown a treasure to dwarf any of the natural ones I had seen on the Trip, and I had just laughed it off as a fake. I guess I'll never know.

Already wise to way India worked, I managed to persuade my thespian guide that I did not want to visit a local market . . . usually code for a sari stall owned by his brother. Instead they took me to see the India Gate, an impressive arch modelled on the Arc de Triomph, and afterwards to another smaller temple where one of the last Mogul emperors was buried.

It was here that I first noticed that in India, unlike in Southeast Asia, the cities were full of animals and birds. This was partly because, even in the cities, the people lived a largely agricultural existence and it was not unusual to see a herd of goats being driven along major roads, traffic coming to a stop to let a cow cross or a group of camels tethered to the railings of some official building.

But there were wild animals as well. Trying to take a photo of the India Gate, I was constantly pestered by a troop of hungry Rhesus monkeys, and it was here I spotted my first Indian bird, a wintering White Wagtail. Later on at the temple, the grounds were full of Asian Pied Starlings feeding on the lawn, and the surrounding palms held noisy flocks of Alexandrine Parakeets. It was a welcome change and it boded well for the cities to come later on my itinerary.

My next journey was a very long one, north out of Delhi and then up into the Himalayan foothills to a National Park called Corbett – after Yellowstone and the Kruger the oldest in the world and named after Jim Corbett, a famous British hunter, explorer and naturalist. This was my first chance to see a tiger, but would also be good for Himalayan bird species wintering at lower altitudes at this time of year.

Today was the day, however, when I learnt that it was not the destinations but the travelling that was the most compelling part of this bewildering country. My driver Divender negotiated our way through the early morning traffic, with many lorries carrying dozens of manual workers into the city, all standing up in the back and grimly holding on to each other for dear life.

We passed several big shantytowns, with whole families living out of a corrugated iron shack, and rubbish tips where the children were already combing the stinking refuse for discarded items to use, eat or sell. Everywhere there seemed to be packs of dogs roaming around, competing with the other animals for living space. The poorest just slept on the pavement, people stepping over them on their journey to work. Delhi, in many ways, was a city from hell – and the choking traffic jams and suffocating heat hadn't even started yet.

Driving in India was a revelation though. There were few road signs and no road markings, so the traffic just made up its own rules based on size.

Top of the tree came the huge lorries, belching black smoke and covered in lurid paintings and religious artefacts. Certainly St Christopher, patron saint of travellers, would have had his work cut out here, as these brutes sped along at whatever speed and side of the road they fancied, and everybody else simply got out of the way.

This rule passed down the food chain. The crowded buses came next, although we saw several turned over on the road, the result of 'king of the highway' face-offs with the trucks. God knows how many people died, but frankly I was surprised to find that some of these crashes still made the papers the following day, as they seemed so commonplace.

Fortunately for me, Divender and I were next in line, in a car, and a jeep at that. After that came the farm carts, pulled by oxen or horses, and often overladen with hay and small children; then the courageous cyclists, who I imagined had long since lost their knees and elbows in the busy traffic; and, finally, the pedestrians. Even in a country this size, and on roads that weren't roads at all never mind had pavements, we found people walking long distances from one place to another in the searing heat.

The protocol on the road was simple – if you wanted to overtake, you waited until there were only vehicles *lower* down the food chain on the opposite side of the road, pulled out and sped past honking loudly and incessantly. The others just got out of the way however they could. This made for a very noisy and scary ride out of Delhi and into the countryside.

The long journey north was a wild adventure through Indian rural life, seemingly little changed in hundreds of years. The Western influence so obvious in Delhi was completely absent here. Most of the surrounding trees had been cut right back for firewood (which did little for the bird life along the road) and every village was signified by at least one huge conical pile of cow dung patties, used for fuel and also as house bricks. Living in shit? You'd better believe it in India.

The north-western part of India had been experiencing a major drought for a couple of years and, consequently, at each and every surviving waterhole there was a hive of activity, with women and young girls balancing vessels of water on their heads. Out here the government's hand seldom stretched, and we passed more than one village where everyone, old and young, were feverishly working to repair their bridge, without any mechanised tools at all.

Stopping briefly for a late breakfast of fried Paneer cheese and black tea, we finally began to climb higher around noon and arrived at my lodge in the mid-afternoon. This was a collection of reasonably well-appointed brick bungalows alongside the riverbank, where the water was reduced to just a stream in the middle of the shingle banks at this time of year.

However, this was more than enough to attract the bird life, and right away I began to sense that here in Corbett the Bird List was going to get a much needed boost. Eating a late lunch, I saw Egyptian Vultures and Black Headed Ibis overhead, with Blue Whistling Thrush and Rose Ringed Parakeet in the gardens and Black Redstart, Grey Wagtail and White Throated Kingfisher along the riverbank. This last bird chalked up number 750 for the Trip, some eleven weeks after bird number 500 in Australia, half a world away.

Just how good India was going to be for birds started to sink in on my first game drive in the late afternoon. We drove into the national park, immediately seeing just how prolific the chital and sambar deer were in the park. Both were primary food sources for the tigers although no one seemed quite sure how many were in the park, with official statistics being widely thought of as exaggerated. In such a vast area as this they were going to be hard to see anyway.

Not so the birds however. Once aboard our elephant (and away from another carrying a party of Indians, who seemed incapable of communicating at anything less than a bellow) we ambled through the bush, connecting with displaying Indian Peafowl (or Peacock to you), Grey Hornbill, Rufous Treepie and Changeable Hawk Eagle, with Bank Mynahs in the settlements. Perhaps the highlight, though, was a snoozing Indian rock python, oblivious to us under a large thorn bush. The last bird of the day was a calling Asian Barred Owlet right outside my room.

Corbett was a country of large (mainly dry) rivers, steep gorges, parched meadows and dense forest, with a sizeable reservoir in the middle. My first full day in the park meant getting up before dawn, putting on lots of warm clothes against the unexpected chill and entering the park with a motley crew of other tourists in a large safari truck.

There were the inevitable delays with permits as we entered the park proper, with legions of officials all wanting something in triplicate, which gave us a few minutes to explore the sadly moth-eaten museum and little concession stalls. Then we were off, kicking up clouds of dust in the now-warming morning air, with the sun rising over the herds of chital deer that were everywhere.

It was on the long drive to the main riverbed in the centre of the park that I found my mind constantly tormented by thoughts of Naazlin. I had called her on the lodge's antiquated phone system early morning her time, only to find that she had just got in from a long hard night partying with her ex-boyfriend in London. Despite her protestations of innocence, I just couldn't get my mind around the fact that she was so loving to me on the phone and yet was obviously not just having a close relationship with her current boyfriend, but now with a former one as well. Despite the glorious morning light and the fragrant Indian landscape in the foothills of the Himalayas, my thoughts were heavy, brooding and focused on a small flat in west London that first morning in Corbett.

Throughout my six months away this was our lowest point, and the only time I seriously considered ending whatever kind of relationship it was that we actually had. If it had not been for the impending meeting in a few weeks in Cape Town, one that I had so carefully set up and nurtured, then that phone call to Naaz may very well have been my last.

My mood gradually brightened as we entered the forest proper, especially when a large raptor flew immediately overhead – a magnificent Crested Serpent Eagle. We walked through the trees to a viewing spot overlooking the river, and the green mists of jealousy finally began to lift from me as I took in the grandeur of the vista, with wooded hills framing the bend in the river, now dry except for a stream in the middle. Below us, our guide pointed out the huge fish in what was left of the water, and there on the bank, snoozing contentedly before their next ready meal, were two species of crocodile – a couple of the usual freshwater variety, but also six of the gharial, a curious beast with a long thin snout especially adapted for eating fish and nothing else.

From here on the day just got better and better. We stopped at the reservoir for lunch and, quickly abandoning the rancid packed lunch from the lodge, began birding the waters edge from quite some distance. In an hour I managed to pick out River Lapwing,

River Tern, Crested Kingfisher, a small flock of Black Storks and a large one of Small Pratincoles, a wader that nevertheless hawked for insects above the water like a swallow.

Having avoided the local troop of grumpy rhesus monkeys, I discovered the local scrub had plenty to offer too. Here was the Himalayan Flameback, the local woodpecker, and favourites familiar from home, a Common Stonechat and Spotted Flycatcher, alongside exotics such as Grey Headed Canary Flycatcher and Dark Throated Thrush (although the last one of these I had seen was a vagrant in a front garden in Maidenhead).

We sped on through the forest, disturbing Kalij Pheasants and the Red Junglefowl – so much more impressive than the farmyard cockerels that were their descendents. Animals were much in evidence, too, as the afternoon began to cool: black-faced langur monkeys atop the fruiting trees, with chital gathering underneath to feed on the spoils; muntjac males barking in the depths of the forest; and family parties of wild boar scuffling around in the dust at the edge of the road.

However, the best was yet to come. We had been following fresh elephant spoor in the road for over an hour when suddenly the bush in front of us moved. We screeched to a halt, and watched it move again, and then slowly we were able to make out the shapes of Indian elephants, wild ones this time, deep in cover. With patience, they moved into view, became bolder and walked out onto the road behind us, with a wary male letting rip at us with an ear-splitting trumpet before leading the herd of some twenty-odd animals to forage afresh in another part of the forest.

Corbett is one of the few places where wild Indian elephants can still be seen, and even here it is getting harder, notwithstanding the National Park status. Despite the profusion of official-looking guards with rifles, poaching is back on the increase here, as evidenced by the fact that I counted only three males in the herd, and all of those were juveniles with only tiny tusks. But at least there were also six baby elephants, which I hope augers well for the future of these dignified denizens of the forest.

The one animal missing from an otherwise perfect day was of course the Bengal tiger. We had seen several tracks and more than once had stopped and waited silently (or as silently as it is possible with a truck full of excited tourists) after the alarm call of a langur or muntjac had raised our hopes. One of these stops had actually been very close to where a British birder, foolish enough to get out of his vehicle and enter the forest, had been killed by a tiger a few years before.

With our minds still buzzing from the encounter with the elephants, and the sun rapidly heading for the horizon, we made our own tracks for the park gates. I had mentally budgeted for only one tiger in my entire time in India, despite visiting several tiger reserves, and so was not disappointed that one had not materialised on Day Two.

The whole truck began to doze in post-safari reverie as we neared the final river crossing before the gate. Then, almost all at once, came three simultaneous sounds. First, the guide shouted 'Tiger! Tiger!'. Then the brakes screeched as we skidded to a halt in the riverbed shingle and, finally, there was a dull 'thump' as my head connected with the back of the cab in front of me.

Shaking off the pain, I stood up with all the others and followed the guide's pointing fingers. In the long grass ahead, something had just moved through, only seen by a handful of us. I was thinking to myself 'just my bloody luck', when off to the right I caught a glimpse of a tiger's head, almost invisible in the grass in the fading light. I was still asking myself whether I could count this as a genuine sighting, based as it was on such a quick look, when from even further right a juvenile tiger emerged onto the bed of the river. For some five seconds it stood there, impossibly handsome in its black, orange and white livery, and looked at us. Then it growled its disapproval and followed its companions into the grass and safety.

Most of our party were struck dumb by this sudden bonanza, although some of the Indians aboard could not resist a very loud 'Oh my God! *Oh my God!*' as the tiger displayed for us. As soon as it had gone, there was uproar in the truck. Several of the women were in tears, everyone was chattering excitedly, and the driver and the guide were happily congratulating themselves; they had not seen a single tiger, never mind a mother and two juveniles, for months.

It was wonderful to see the guides as happy as we were to see these awesome beats, which immediately relegated everything else I had seen on the Trip to second best. And it was doubly satisfying knowing now that I could afford to relax a little having seen my first tigers so early in the trip.

It was a very smug party that returned to the lodge, and began to regale the less fortunate fellow guests with the story of *our* tigers. After an inauspicious start, today had gone on to become one of the best days of the entire Trip so far. It also made that evening's dinner a little more palatable.

I had been looking forward to eating authentic Indian food on this trip, and was relying on my cast-iron constitution to see me through the legendary digestive problems. Many people I knew who had been travelling in India avoided the meat dishes, and originally I thought that this was because this was the best way of avoiding 'Delhi Belly'.

However, I was now in the process of finding out the real reason. Forbidden to eat both beef and pork for religious reasons (one is holy to Hindus, the other unclean for Muslims), and with no access to the sea for seafood or fish, all the meals here in the north of India seemed to consist of either mutton, swimming in grease and with glutinous bits of white gristle attached, or chicken, usually on the bone with maximum skin and minimum meat content. Certainly no chicken tikka masala here. It was curry for breakfast, lunch and dinner.

Many of the guests at the lodge were experienced travellers, and I found that my tales of the last few months were by no means a surprise to many. Birders were much in evidence, and I began to wish that I could follow them higher into the mountains later in the week in search of new species. I decided to hook up with another birder and hire one of the guides for an early morning bird walk along the river and into the hills the following morning as the next best thing.

At this altitude the early morning was downright cold, in stark contrast to the heat in the middle of the day. We stood and shivered in the hotel grounds, nonetheless finding warm enough hands to tick the Spangled Drongo, with it's ridiculous long feathers on its crown, and the locally uncommon Blossom Headed Parakeet.

We crossed the river further up, having seen both White Capped and Plumbeous Water Redstarts, and Brown Dipper, a pair of which were 'dipping' in a small set of rapids by the bridge. Further up, there was a very impressive Hindu temple on top of a tall pinnacle of rock in the middle of the river, an obvious photo opportunity in the pink morning light.

Crossing the dry riverbed at this point, we moved into the scrub on the far side, the air full of calls from wintering Chiffchaffs (of the Siberian sub-species) and Plain Martins. Here was a profusion of birds for us to gorge on – four species of Bulbul alone, including Himalayan, plus both Asian Brown and Rusty Tailed Flycatchers, a beautiful metallic blue Small Niltava and the distinctively gaudy Emerald Dove.

The bird count mounted as the sun rose and the habitat changed once more, and we climbed higher into the forest proper. This was a chance for us to test our identification skills on one of the most difficult families of birds, the Leaf Warblers – to a bird, all small, green and mobile, which usually meant you had a split second to look for the relevant eye stripe or wing bar before it disappeared back up into the canopy.

Nevertheless we were sufficiently sure of out identification to log Brook's, Blythe's and Hume's Leaf Warblers, paying our own homage to the naturalists of yesteryear after whom these were named. However, my favourite bird from this trip was either the Black Lored Tit, which looks like our own Crested Tit except that it has been dipped in bright yellow paint, or the Brown Fish Owl, roosting on tree branch with the nearest water a mile or so away.

By midday I was beginning to realise that my fitness was not up to this marathon, especially without a stop for water or rest. By now the birdlife had begin to quieten down, although a distant Grey Winged Blackbird almost made up for us missing the extraordinary Forktails that were a target species in this part of the forest. Finally, we began to descend the far side of the hill, and through the trees you could see the bright green of the cornfields surrounding the huddled village below. Some half an hour later, I was glad to be out on level ground and in the sunshine once again, watching the curious children running home from school dressed in their best clothes and their parents proudly tending their fields.

We were finally picked by our driver at the road, still another mile distant, and I was grateful to collapse into the back and snooze all the way back down to the lodge. It had been a forty-six-tick day, the best for three months to the day since the rainforest in Ecuador. This one day alone had yielded three times the Vietnamese total, or nearly as many as the totals for Thailand or New Zealand. Suddenly that 1000 bird total was looking within reach again.

My last day at Corbett began with another dawn jeep ride through the park, where we saw tiger and sloth bear tracks, and civet droppings, but no signs of the animals themselves. Instead, the best find of the morning was a fine Crested Tree Swift, elegantly flicking its way across the tree line.

With the afternoon to myself, I thought I would while away the hottest part of the day with a book by the pool, but hadn't allowed for a gaggle of Indian teenagers seemingly going through puberty in front of my eyes, and with a wider selection of English swearwords than I had ever heard in London.

This break gave me a chance to reflect a little. In just over a month I would be back in chilly, foot-and-mouth-bound UK, and yet in many ways it seemed I had only just begun to adjust to travelling by myself abroad. Those first few angst-filled weeks in the USA already seemed like ancient history, and although it was only two months earlier that I had seen Naaz in London, that too seemed like years ago. For the first time, I began to worry what would happen when I returned home, jobless.

The best antidote to all this introspection was a final bird walk along the river in the calm of the afternoon, which yielded a fine Wallcreeper amongst the boulders. This species is much sought after in Europe, with people travelling to the French Alps or Spanish Pyrenees to see it, but here I had one at close quarters within sight of my hotel room. The other golden tick of the afternoon was back in the scrub by the temple we had discovered yesterday. On that occasion we had found a White Tailed Rubythroat alongside a small pool, and today I was able to add its cousin from the Russian taiga, the Siberian Rubythroat, to my List, from a similar site nearby.

Corbett had been a revelation in birding terms, with exactly 100 new species putting me firmly back in the driving seat. The only disappointment was not seeing the Ibisbill, a colourful wader which breeds in the high mountains but which winters at these latitudes. It had departed for its breeding grounds days before we got there. Much would now depend on whether the other reserves further south could add sufficient new birds to give me a healthy lead going to South Africa for the last leg. But, of course, the real highlight didn't have a beak or feathers – that first sighting of a tiger would stay with me for a very long time.

Divender appeared the following morning, looking a bit sheepish; although I had allowed him to travel the extra 50 kilometres home rather than stay with the other drivers at the lodge, his sense of duty still meant that he was very apologetic. It seemed the only fair thing to do as he only made it home on a handful of occasions during the tourist season. He had arrived fresh from the family temple, with a bright yellow *bindi* painted on his forehead and leaves woven into his hair at the back.

I dug in for another day-long journey south, this time beyond Delhi to the bird sanctuary near Bharatpur in parched Rajasthan. Once again this was not wasted time, since there was always a street drama outside my window to entertain me, and also because it enabled me periodically to get a signal on the mobile phone. Perhaps the most surreal episode of many in India was to be stuck in a traffic jam in some obscure provincial town, surrounded by honking taxis, sweating oxen and staring peasants, and then giving Divender my phone so that he could chat to Naaz in London in Hindi. It was a nice warm feeling to be told by my driver in broken English that Naaz had asked him to take special care of me.

Bharatpur was nine hours distant, nine hours of dusty, pot-holed roads, countless naked ragamuffin children and just one stop for a cup of black tea. I amused myself with a mental checklist of the colours of the women's saris – however poor they seemed to be, every woman in every village was resplendent in her brightly coloured garb. Not surprisingly perhaps, the most popular saris were green and orange, the national colours.

When we finally arrived at the Forest Lodge, the only hotel actually inside the Bharatpur reserve, it was too late to get any birding in and relieve the atrophy in my legs. The lodge was a rude shock after Corbett; run by the government, it was apparent very early on that the staff weren't exactly motivated, and the building itself was reminiscent of a third-class Mediterranean motel from the 1960s. Its only merit was that the riches of Bharatpur were right outside.

And so it was that very early the next morning that I hooked up with Vivec, my guide at Bharatpur, and Ashrampouri, my rickshaw driver. They were as different as chalk and cheese.

Vivec was not a birder, although did have a modicum of knowledge about the park. He was a round faced, clean cut, twenty-something, desperate to please and the kind of correct young middle-class Indian that was probably the apple of his parent's eye. Unfortunately he could never remember my name, and for four days I was 'Mr Johnson'. I soon gave up trying to correct him.

Ash, on the other hand, was a thorough rogue. He had the haunted eyes, filthy rags and piratical moustache that suggested that if he were not making a living ferrying birders around the park, he could easily pass for a street pickpocket. But he had a keen eye for birds and the three of us prepared to milk Bharatpur dry.

The park, like the rest of Rajasthan, had not seen rain for two years, and two-thirds of it was dry and dust-bound. The remainder was being irrigated artificially from precious underwater springs, which at least had the effect of giving enough surface water to sustain part of the park in the style that it had become famous for.

The wet part of the park consisted of several large pools, dotted with islands and surrounded by raised banks for walking and viewing. In former times, this whole area was a hunting preserve for the maharajas, and a tablet in the centre logged their daily 'bags' – sometimes more than tens of thousands of ducks daily in the 1930s. It made sickening reading some seventy years on.

Numbers were well down these days, not least due to the reduced water available, but I still hoped to get most of the key species with a little help from my Indian friends. However, one bird I already knew I wouldn't get was the Siberian Crane. This was an increasingly scarce species from Russia, which had steadily been declining for decades in the face of persistent hunting and the conversion of its wetland wintering grounds to agriculture. For some time Bharatpur has been the last remaining regular wintering site, and in the last few years numbers have been pitifully low. The only two birds seen this year had both flown back north, to an uncertain future, nine days before I got there.

However, Bharatpur was still to offer me an orgy of bird-watching that day, exactly one month from going home. It was the second most prolific day (after one in Ecuador) of my entire journey.

The wetlands, small as they were, were still stuffed with water birds – residents like Bar-Headed Goose, Bronze Winged Jacana, Spot Billed Duck and Indian Cormorants, local migrants like the White Tailed Lapwing, plus birds from Europe like Eurasian Wigeon, Teal and Ruddy Shelduck.

As well as waterfowl, there were innumerable others relying on the water as a food source. Although the Siberian Cranes had gone, there were still flocks of both Sarus and Common Cranes on the mudflats, illuminating the flatlands with their evocative honking as darkness fell. There were Painted Storks, the quintessential bird of Bharatpur, but only here in dozens rather than hundreds due to the low water levels and their breeding failure that year. There was also the elusive Black Bittern that Vivec managed to find for me amongst the dense reeds, a dozen or so species of wader from Northern Europe which presented no identification problems at all and, finally, the Clamorous Reed Warbler, the most surprising discovery of all in a tree by the edge of a canal.

Bharatpur was also a haven for raptors and vultures, attracted both by the food offered up by pools that were drying out, and the large amount of both domestic and wild livestock grazing on the reserve. The commonest bird of prey was the Lesser Spotted Eagle, which seemed to thrive on the mixture of wooded islands and open water, and its cousin, the Greater, was also around in smaller numbers. However, the highlight of the day was the totally unexpected sight of an Imperial Eagle scuffling on the ground with a Tawny Eagle, trying to dispossess him of a snake for that night's dinner – and all just 10 yards away on the other side of a creek.

Later we stumbled upon a dead cow in a distant field, and with the bloated corpse were a large mixed flock of vultures – the Long Billed and White Rumped ate first, whilst the more numerous but smaller Egyptians waited their turn at the table from a distance.

This was all the more exciting as the whole of India's vulture population was in the process of being decimated by a mystery virus, which partly explained the growing number of wild dogs that had taken their place in the food chain.

The next couple of days were spent with Vivec and Ash assiduously covering every inch of the reserve. They took me into the temple grounds, where I avoided the curious gaze of the monks and a gaggle of local villagers to pick out Coppersmith Barbet and several superb Orange Headed Thrushes. Then we spent some time stalking a colony of Eurasian Thick Knees (or Stone Curlews as we know them), at the same time scoring with the scarce White Capped Bunting, and a regal looking Scaly Thrush on top of a distant tree. Vivec's moment of triumph came when he took me to a thorn bush in the middle of a meadow and showed me the nest of the Dusky Eagle Owl, complete with two downy and curious chicks.

There were simply birds everywhere. Look up, and there would be an Oriental Honey Buzzard flying over or a Shikra chasing small birds would flash by. Sit down for a drink and a Dusky Warbler would come and bathe at your feet – in Europe the most skulking and elusive of birds, but here as tame as you like. Look out to the lakes and there was a circling flock of immense Great White Pelicans, several unfortunately trailing lengths of fishing line attached to their legs. Go back to the hotel, and there would be an Olive Backed Pipit in the brush outside my window. After such a long absence here in Asia, the glut of birds in Bharatpur was like a life raft to a drowning man.

On my last full day at the reserve, Vivec humbly asked me if he could have the day off to attend *Holi*, a Hindu festival honouring the God Krishna. I had been seeing carts in the streets carrying huge piles of brightly coloured powders for days, and this was what it was all for – the festival involved liberal use of coloured dyes, sprayed over anything and everything. Of course I said yes, especially when he offered to take me as well.

Vivec was to celebrate *Holi* at the hotel owned by his friend Dipraj, an old colonial villa outside the park now lovingly restored. Vivec arranged to pick me up on his scooter, and I prepared to wear my oldest clothes for the occasion – partly because I knew that foreigners were usually the most likely targets for the technicolour onslaught, but also partly because my constitution had finally given in to the horrible food in the lodge, and I was not sure whether a nasty accident was on its way.

Much to the amusement of my fellow guests, most of whom were sheltering in the lodge for the day whilst the country virtually shut down, Vivec arrived on his scooter at the allotted time, covered from head to foot in bright purple dye. We managed to negotiate the gaggle of small boys armed with water pistols on the way to the hotel, but once inside I was handed a beer and then smeared with dark green paint almost simultaneously.

The next few hours were a delight, being amongst a local community (albeit a rather wealthy one) and being part of one of their biggest festivals. The local women, as is their custom, arrived to dance and sing for us, throwing great clouds of powdered dye in the air. The more adventurous of us got close to join in, whilst the more timid stayed back and viewed from afar. Either way, by lunchtime most of us were of a rather lurid hue, and I discovered that at least in Dipraj's hotel, Indian food could be delicious after all. It was a real privilege to have been invited to this most Indian of gatherings, although with my guts in mind I declined Vivec's kind invitation to carry on the celebrations at his own home. Just as well, for it took me the rest of the day to get the dye off, and I was still sporting pink fingernails in South Africa weeks later.

The lodge may have been rather tatty, but it attracted some fascinating clientele. There were from England three birders of some eminence, who got me salivating over their stories of the birds and tigers still to come in the reserves I was to visit next; the British Army Ornithological Society, who were always terribly present and correct, and were planning their birding expedition like a military exercise (their holiday was actually called Operation Black Eagle) and, finally, there was Pat and his Norwegian wife.

Pat was fascinating character. He had spent most of his adult life in the Royal Marines, and been in the SAS at various times. White haired and bearded, he had dark eyes that were still full of adventure and life, and I admired the fact that even in retirement he was still as active as he was in the services. He had been on several madcap expeditions, dreamt up by ex-Army colleagues of his, to map the upper reaches of far-flung rivers in South America and to raft down rapids in Africa. 'I hope I can be as adventurous as Pat when I'm his age' I found myself thinking, as if to assume that at *this* age I was already – which was doubtful.

Divender duly appeared on my final morning, still covered in coloured paint from his own *Holi* celebrations, which he had had to hold with the other drivers at the lodge rather than at his own home. I bade a fond farewell to Vivec and then Ashrampouri, and a rather less fond farewell to the lodge staff who had suddenly appeared to wave me off – all expecting a tip which didn't materialise. If only they had been so bloody attentive when I was their guest – although I do have to give credit to the manager, who I pestered relentlessly in order to use his office phone to ring Naaz in London.

Bharatpur had added another sixty-three species to the impressive total from Corbett, and now that list was back on track, having passed the 900 mark. Another 100 to go, I thought, and just as well as my next stop was definitely not of the birding type. I was to become a tourist for the day and fulfil another long-held ambition – to visit Agra and the Taj Mahal.

BIRD COUNT: 905

12 A Tiger Betrayed: Bandhavgarh and Kanha, India, March

It was only a couple of hours by road to Agra, but in India this is time enough to be fascinated by events right outside your car door. First, we had swerve to avoid several crowded mini-buses full of revellers obviously still drunk from their *Holi* celebrations the night before; no sooner had we got rid of them, than we had to drive past dozens of street vendors, each with a poor sloth bear tethered to a pole by its nose, and goaded into dancing for the Western tourists. If I could have persuaded the drunken drivers to aim for the bear owners I would have been much happier. Life of all kinds is so cheap in India.

Later, on the outskirts of Agra, we stopped to let a wedding procession pass on elephant back, and whilst we were stationary two Indian teenagers thought it would be fun to yank open my door and spray the tourist inside with purple dye. Only Divender's sharp rebuke saw them off before any damage was done, although I suspect he was just as worried about his nice white jeep as his nice white tourist inside.

And then, finally, in the centre of Agra, we got caught up in a huge gridlock for an hour. Amidst much shouting and gesticulating, the traffic only got going again when a Police Commissioner waded into the crush and started to lash out with a small hand whip like a riding crop. The fact that the Police Commissioner was stained purple from head to foot lent the occasion a more than farcical air.

I finally bid goodbye to the trusty Divender, who had been stuck with me for over a week, and for several hundred miles. A quick change and I was off to the Taj, keen to see it at its best in the late afternoon sunlight.

The Taj Mahal proved to be one of those rare sights that is *so* impressive that it transcends the hype – it is actually even more beautiful than you think it is going to be, a quality it shares with the tigers of India.

It was built in 1631 by Shah Jahan, one of the most powerful of the Mogul emperors, to mourn the death of Mumtaz Mahal, his favourite wife, as she gave birth to his *fourteenth* child. I couldn't help but think she was grateful for the rest. By the time it was finished twenty-two years later, it had employed 20,000 workers, the best stonemasons and artists from the known world, and all the money in the royal coffers. But you have to say it was worth it.

It's hard to forget that first sight of the Taj – for me it was from my hotel window, and the dome and its attendant minarets rose high above the low-rise shanties of the centre of Agra like a jewel in the dust.

As we drove nearer, we were obliged to leave the car behind and use a diesel bus, in order to avoid adding to the local pollution, although the marble walls of the Taj looked unblemished by the twentieth century. Passing through the red sandstone gates (impressive in their own right) you are faced with *that* view, looking down across the gardens and the fountains towards the entrance to the Taj itself. You half expected Princess Diana to emerge stage right, as it was such a famous image the world over and one she had done her best to instil in people's minds.

The Taj was crowded with weekend tourists, all taking the same shots as me but all of us richer for having it on film. It had a mesmeric presence, and you felt compelled to just sit and watch the marble slowly change hues as the afternoon waned. Tearing myself away from the view, I meandered down the avenue towards the Taj, trying somehow to stare so hard that it would always be imprinted on my brain. As you got nearer, so the immaculate carving around each door and window became evident. Each and every detail had been thought of, with the marble keeping the insides cool to the touch even in the hottest conditions. Around the main door were extracts from the Koran, carved into the marble and inlaid with semi-precious stones, whilst some of the filigree carvings by the Persian stonemasons, from one single block of marble, were simply exquisite.

I looked out over the riverbed behind the Taj, and tried to imagine the barges bringing their cargo of stone from Northern India right up to the walls, with slaves then toiling to bring the blocks up the hill to the construction itself. How much blood, sweat and tears must have gone into this magnificent building I thought.

The inside was the only part of the Taj that proved disappointing. Surprisingly low key, the emperor had perhaps wisely not gone for thousands of rubies or emeralds, for they would surely have been looted by now. The stonework, itself inlaid with designs edged in semi-precious stones, said everything he needed to say about the love he had for Mumtaz.

Outside once more, an enterprising priest showed me a selection of classic camera angles to use up yet more film on the Taj, all in return, of course, for his 'thank you'. It was worth it just to capture it at its very best, as the white marble changed from white to ochre to rose pink as evening approached.

And by no means all the attractions here were made by humans. Above the Taj, flocks of Pariah (Black) Kites soared, accompanied by several Egyptian Vultures, species that had probably inhabited the building since it was erected. Around the outbuildings I found nesting Dusky Crag Martins, House Swifts and Rose Ringed Parakeets, all adding their noisy tribute to the grandeur of the Taj. After an hour, it was all I could do to tear myself away – I somehow felt that I should stay there until dark, as I was never likely to see this place again. Unfortunately the security situation meant that it was no longer open at night, although I could see it resting majestically in the moonlight from my hotel window. How I wish Naaz had been there to see all this with me – a kiss in the moonlight with the Taj as a backdrop – does it get any more romantic than that?

The next day was to be spent in Agra, and its gentle pace was welcome with so much travelling still to come. A visit to Agra Fort gave an insight into how both the Moguls and the British lived here, but I found it chiefly memorable for its views of the Taj from across the river.

The afternoon was spent just chilling – a badly needed haircut, a gloriously cheesy pizza (being bored with Indian food after just ten days in the country), and a bizarre customer service call from my mobile company in the UK ('yes I'd love to talk about your new tariffs right now, but I'm here looking at the Taj Mahal at the moment') passed the afternoon. I wanted to walk around the town centre but, as in other Indian cities, I found that there were no pavements, no centre, too much traffic and too much hassle as soon as you stepped outside the door. I became almost a prisoner in my own hotel.

I used the time to ring home; James was cock-a-hoop at Spurs' latest victory, whilst Mum and Dad were as concerned with all the minutiae of life – laundry, diarrhoea, loneliness, money – as with my stories of the Taj, bless them. How could I expect them to be as impressed as me? You just could not describe that feeling to someone who hasn't seen it. And Naaz seemed finally to be coming to Cape Town, having begun to work on her boyfriend to make it as easy on him as possible.

The next day began with a quick hop by air down to a town called Khajuraho, where the following day I would pick up my car and driver and begin our long trek to the tiger reserves in Madhya Pradesh in central India. I wasn't expecting much from the town except a free afternoon, so to find that it was a sleepy and charming little market town was a nice surprise after all the hustle and hassle of the other Indian cities I had been to.

The town's chief claim to fame was its temple complex, some of which dated back to the tenth century. This made them three times as old as the Taj Mahal, and the same age as the temples in Sukhothai, in Thailand, but much better preserved (well, they had never been looted).

I am always amazed that such fine architecture and deep-seated culture can be so easily lost, and the world is full of examples of cities and temples becoming dilapidated and overgrown over the centuries, only to be rediscovered much later, usually with Western help. The Khajuraho temple complex was a good example, being 1000 years old and 'lost' for much of that time, until fairly recently.

The temples were truly impressive, covered as they were in yellow ribbons and purple bougainvillea, not to mention the extraordinary carvings that covered every square inch of space.

Unlike most religions, the Hindu faith celebrates the sexual act, and as such most of the temples here had carvings depicting just about every sexual activity I had ever heard of, plus a few I hadn't. It seemed that the whole of the *Kama Sutra* was portrayed here, and I was particularly impressed by the way that the single man in these friezes often recruited the help of his lover's hand maidens in helping her to perform the more gymnastic of sexual positions. All very entertaining, but to a man who had been on the road for five months, and barely even *seen* a woman for two weeks, it was also a rather thought-provoking afternoon shall we say.

I met my new driver the following morning, a rather decrepit, wizened old man in a similarly decrepit and wizened old car. The vehicle I was not worried about, since I had learnt that the Ambassador car, the vehicle that held the Indian economy together, was virtually indestructible. I was less convinced of the durability of the driver, especially when I heard that he had already driven for four hours to get to me from Bhopal, and that the road ahead would be far worse than anything I had encountered up until then. I quietly wondered if we would see my next stop, the tiger reserve at Bandhavgarh, before dark, some ten hours later.

The road to Bandhavgarh was not so much filled with holes, as holes interspersed with bits of road; for the remainder of the day we averaged just 20 miles an hour, much of it spent carefully driving in and out of huge potholes, or along the verge when the tarmac just stopped.

I marvelled that our ancient Ambassador survived the constant buffeting on these dirt tracks, and that my driver stayed awake in the constant rocking and rolling, like a rodeo in slow motion. Only once did I even see him take a drink of water.

The usual cavalcade of sights, sounds and smells of India stopped me getting too bored. This much further south, the locals were burnt dark by the sun, and the extreme poverty of the rural population in this state was both fascinating and at the same time disturbing. The children were what I remember most, almost all naked and filthy, yet still adorable. They thought nothing of coming right up to the car window and staring in at me like some animal in a zoo – its funny how you can so quickly get used to the array of big brown eyes at point-blank range.

After a couple of hours, as we began to climb into the hills that dot the entire centre of this part of India, we came across the first of two huge gypsy caravans. I already knew that the gypsies we see in Europe were meant to have originated in India and, looking at the faces of those aboard the wooden carts ahead, you could see the similarity with the swarthy gypsies we see at home. However, there the likeness stopped.

This caravan consisted of some fifty carts, pulled mainly by oxen and usually driven by small boys. The men rode separately on horseback, with gaggles of womenfolk congregating in the bales of hay or on the piles of wooden furniture inside the cart. The women were adorned with bangles and necklaces of gold but, that apart, everyone wore clothes that looked as though they would already have been old years ago. Alongside the carts trotted the usual pack of ravenous dogs, whilst tethered to the rear of most was a weary looking cow. How these rickety vehicles ever made it down the hill on this poor excuse for a road was beyond me. I was left thinking that these noble and distinctive people seemed to have little in common with the social outcasts we know in Europe. I wonder how many of those though, if they were able, would exchange a caravan park in Sidcup for this life in India.

We stopped just once, in the middle of nowhere, where suddenly a shack had sprung up on the side of the road, with the usual array of pots and pans around an open fire, and beds set up along the road for weary drivers. With the large distances involved and non-existent roads, I assumed that many drivers must be away from home for days at a time.

I bought my driver some tea, *chapati* and *dhall* for less than the price of a newspaper back home, and wandered over the ridge to view a distant reservoir from the crest of a hill. There were hundreds of both Pochard and Tufted Ducks present, birds I was used to seeing on local windswept reservoirs in chilly West London, and looking incongruous in such warm and exotic climes. On the way back to the car I was able to keep the List ticking over with a Large Grey Babbler, Rufous Tailed Lark and Crested Bunting.

It was mid-afternoon before I reached my lodge on the edge of the national park. By this time my buttocks were so numb that I couldn't feel a thing – just as well since I immediately went out with a driver and guide in a jeep, and spent another couple of hours on a metal seat haring around the dirt roads of Bandhavgarh. Incredibly, my driver apparently slept for a couple of hours, and then turned around and did the whole journey in reverse – and this time in the dark. Perhaps he was really a young man, who only *looked* old as a result of having to do this trip on a regular basis.

Bandhavgarh was different to Corbett, being less mountainous and less affected by drought. It was criss-crossed by wooded streams and meadows, with a large rocky escarpment in the middle, but had the significant advantage of a healthy breeding population of tigers in a relatively small space, making them easier to find and observe. Unfortunately, this being India, this fact had not gone unnoticed by local poachers, and every year a few tigers fell victim to their guns, including a famous local tigress who had given birth to fourteen cubs before being killed only a year or so ago.

That first game drive proved fruitless, and we returned to the camp and a well-earned rest. The other guests were an interesting bunch – two very large English ladies indulging their passion for exotic travel and wildlife; Bart, an eccentric character and the world's only humorous Belgian; a party of noisy and drunken Indian travel agents, on a junket from Delhi; and last of all Richard Newton, travel and wildlife writer extraordinaire.

Richard had my dream job, travelling the world and writing for newspapers and magazines about countries and their wildlife. Despite the fact that he was freelance and therefore had to hustle for a living, I was envious of the freedom this gave him, and that he was the only person I had met who could manage to make a living from his hobby. He had just arrived after a similarly exhausting journey, and we bonded immediately. Those long dusty journeys in the back of the jeep later on were regularly enlivened by his tales of feeding Tasmanian Devils in Australia, or visiting the Ismaili tribes (from where Naaz's family originated) in the mountains of northern Pakistan.

However, the major entertainment that night, and unbidden at that, came not from the guests but from the staff. I should preface this story by pointing out that the all-male staff were often away from home and loved ones for many weeks at a time, so perhaps I should not have been surprised when, after I had retired to my bungalow and the generators had been switched off, I was awoken my some strange noises from next door.

The manager's bungalow shared the other half of my block, and I was disturbed by sounds of at least two men talking in not-so-hushed tones around midnight. Then it subsided somewhat, but just as I was dropping back off to sleep the talking was replaced by the unmistakable sounds of carnal pleasure. For the next half an hour or so my evening was given an (extremely) rude awakening by the wailing and long drawn out moans of at least two men doing what the Indians quaintly call 'making the beast with two backs'. Now, I'm no prude, but I must confess that it was a strange and unexpected end to a very long day in the car – and I suppose it was not only *my* backside that was a little tender that night after all.

The next day was memorable for both right and wrong reasons. Up as usual at dawn, we were at the gates of the park by 6 a.m., and assigned our guide for the morning. We then spent an hour or so driving around the park looking for tigers, knowing that the mahouts on elephant back had the best chance of finding them, high above the grass as they were, and able to manoeuvre close to the animals without them taking flight – for some reason, the tigers never equated humans on elephant back as humans at all.

Sure enough, by 8 a.m. we had congregated with half a dozen other jeeps at a spot where the elephants had tracked down a resting tigress with two cubs, and the news radioed into base. It seemed an age waiting for our turn, whilst the park officials arrived and, with rare organisational skills, sorted out payment from each of us for our time on the elephant. This was evidently a very lucrative business.

When our turn came, Richard and I clambered on to the roll bar of the jeep and heaved ourselves onto the elephant, using the ropes holding the wooden seat to the animal, and tried to sit with equal weight displacement in the seat so that the poor elephant could walk comfortably. Its rolling gait, leathery skin and puffing progress were all very different to being in a motor vehicle, but somehow much more akin to the natural surroundings, much of which the elephant was trying to consume as we crashed almost delicately through the bush as only an elephant can.

We were almost on top of the tigers before I saw them, their incredible camouflage meaning that it was not until the tigress moved her head that I spotted her at all. She was sitting majestically in the middle of the meadow only 10 yards away, her two cubs peering curiously through the grass from behind her. After the initial flurry of 'Oh my God's' and the volley of camera shots, we had a few moments to take in the spectacle – the animals were totally happy with our intrusion, were always in control of the show and looked at us rather imperiously as though they knew it. Their colours were a sheer joy to behold, with the amber back offset by the white face and stomach, all etched in black stripes. It was at this precise moment I knew that I would not see another animal or even a bird that could compare to this first point-blank encounter with a tiger.

I wish I could say that I enjoyed my second sighting that day as much. The elephants rested in the afternoon, and so my game drive after lunch turned out to be with just the driver, guide and I. I was pessimistic about seeing another tiger without the help of the *mahouts*, and thought as we passed through the park gates, with the huge sign saying that the tiger's safety and protection was paramount and that he would see you but you might not see him, that I would have to wait until the morning for my next fix.

My doubts were confirmed as we spent an hour looking unsuccessfully for tigers, and the sun began to get lower in the western sky. Then, as we drove alongside a meadow, we heard the alarm call of a langur monkey atop a large tree, usually the first sign that a tiger is in the long grass below.

We stopped beside another jeep, and waited for any movement in the waist-high grass. My guide explained that this tigress also had cubs but, unlike the others was very wary and nervous of humans, especially in jeeps, and was rarely seen. Eventually the other jeep left, leaving just the three of us staring at a seemingly empty meadow.

The first I knew that something was wrong was when the guide mumbled something to me and got out of the jeep, a cardinal sin in the reserve for obvious reasons. Out of the corner of my eye I saw him pick something up off the ground, and before I knew it he had thrown a large rock into the middle of the meadow. It kicked up a cloud of dust, but nothing else, and so he tried another, again without success.

I had got used to trusting my guide as the guardian of these tigers, certainly against the myriad of forces lined up against them – overpopulation, poaching, pollution, politics. I was till wondering how he could act so irresponsibly when the driver began to reverse the jeep, with me in the rear to the fore, straight into the meadow.

At this point my emotions were very mixed. Undeniably excited, I was also appalled that these two, who should know better if only because their livelihood depended upon not disturbing these magnificent creatures, would go to these lengths to show me a tiger. I was about to protest, when all hell broke loose.

With us about 50 yards into the meadow, there was a loud roar from the grass in front of us. From my vantage point standing up in the back of the reversing jeep, I could make out two tiger cubs running out of the meadow on the far side and into the safety of the trees, and at the very same moment, a tigress charging the jeep.

With the driver and guide laughing and chattering excitedly like schoolchildren, and me screaming 'Back, back, get back !' we began to drive out of the meadow at speed, as she paused only 20 yards away and roared angrily at us. My heart was beating so fast that I thought it would burst, but as we paused at the edge of the meadow – for the driver to cover up the tracks the jeep had made – my exhilaration turned to anger.

The driver and guide were still full of adrenaline. Had I seen the cubs escaping? Did I hear the tiger roar? Did I get a good close-up view? I knew what all this meant – a bigger tip for them. It was customary to tip the guide and driver if a tiger was spotted (likewise the *mahout*) and they had broken all the rules to extort money from me. But the worse thing was I let them do it.

I tried to make them understand that, far from being happy about that, I was angry, and that they must not do it again, but I don't think they understood. Conscious of the fact that if you did not tip then you got no service the next time around, I even gave them both a tip at the end of the drive, whilst they implored me not to tell anyone. And, even though at the time I promised myself I would shop them, I never did.

In doing this I left Bandhavgarh knowing that I had betrayed the tigers as much as they had. If I had reported them, then perhaps the other guides would have taken a more responsible attitude. But I was too scared that, as had been the case for other tourists, I would get a bad name and not be shown another tiger if the guides could help it. I resented them profoundly for giving me such a terrible choice.

I was to see another mother with cubs the following day, feeding on a wild boar, and finally a superb female again walking serenely across a meadow in full view of dozens of cameras. But I was never quite sure whether those experiences were worth me betraying the tigers as surely as if I had driven the jeep into the meadow myself.

As we drove away from Bandhavgarh, at least I could use the fact that it had provided the usual gamut of new birds to take the nasty taste out of my mouth. The highlights included a flock of sixteen Malabar Pied Hornbills, a couple of Black Ibis at a local dam, two displaying Red Necked Falcons (spring was approaching it seemed in the northern hemisphere), a Lesser Adjutant pointed out by Richard at the local watering hole and an Ultramarine Flycatcher around the lodge. I also missed a wolf that frequented the camp one afternoon when I was in the park. You can't have it all, I said to myself.

The final stop was Kanha, another notable tiger reserve, and the setting for Kipling's *Jungle Book*. Its tigers were already famous through a documentary by David Attenborough (the other man, besides Richard, to have my dream job), and I hoped that in this park at least the rangers would act in the best interests of the animals. Some hope.

The journey to Kanha was a pleasant change – the car was air-conditioned, the roads were in large part intact and everywhere I looked the locals were dolled up in their Sunday finery. The camp, too, was noticeably less mosquito infested than some of the others, despite being on the banks of a river, and in Mr Eric, the camp director, I finally met an Indian with an interest in birds.

My guide, Mr Desmond, was an odd cove though. Well into his sixties, he continued to work as a guide through boredom I suspected – all his family had moved away to live in England or the USA, and he seemed desperately lonely. Despite his obviously

Indian face and nut-brown complexion, Desmond was adamant that he was different from the rest of the staff – he was an 'Anglo' he told me, with white blood and he was a worshipping Christian. He obviously loved the park and its animals, although he was getting a little rickety for dawn starts and long jeep journeys, and his bird identification had not improved with age. On my first night in Kanha he said to me that 'India was a wonderful country, except for the bloody Indians'. I couldn't have agreed more.

Kanha was different again to the other parks. The mixture of wooded copses and shady glades reminded me a lot of the New Forest, and certainly was not what we in Europe would think of as 'jungle'. In fact the only thing all the parks had in common was the crippling bureaucracy in order to enter, and the stress of knowing who and when to tip once you were there.

Our first dawn game drive took us into the centre of the park, where there was a surprisingly modern reception centre, although the hospitality was somewhat muted by all the toilets being infested with huge swarms of wasps. It also possessed a large verdant lake, where I could tick Lesser Whistling Duck, as well as Large Billed Crow, Common Iora and Asian Koel.

We waited there like the worst kind of eco-tourist, munching on our picnic breakfasts with the well-appointed occupants of another dozen jeeps, all waiting for the radio call that said the mahouts had found a tiger. This was the most civilised of wildernesses, I thought to myself.

Then there was a sudden flurry of activity, as word came through that a pack of Indian wild dogs had been found on a kill, and only minutes away from the centre. The convoy of jeeps was soon screeching to a dusty halt about a mile away, and sure enough on the other side of the clearing a pack of dogs, looking like Alsatians with vicious oversized dentures, were tearing into a freshly killed chital.

We watched mesmerised as the strong social structure amongst these animals displayed itself for us to see. Whilst three or four dogs would take turns to scoff the rapidly diminishing corpse (do predators ever get indigestion? This is obviously where the term 'to wolf down' originated), two others would stand guard on the edge of the kill against the growing numbers of Vultures that were suddenly appearing from all directions. It was hilarious to watch the sentinels repeatedly charge the groups of birds, even leaping into the air to try and bring them down – a waste of time, for seconds after they had scattered one flock, they regrouped somewhere else and once again began a stealthy creep towards what was left of the deer.

The most poignant part of the show, however, was the remaining dogs which ferried food back to the rest of the pack; a couple of hundred yards away, half a dozen cubs were hidden in the longer grass, and the dogs took turns to return and regurgitate meat for them. It was a memorable sight and one we all felt privileged to witness.

This being India, however, scandal was only just around the corner. A few minutes after we arrived, a young *mahout* appeared on his animal, wielding an enormous telephoto lens. This was the eldest son of the head *mahout*, who had tragically lost his younger son the previous year when the boy slipped off an elephant and onto the ground in front of a young male tiger. He was already dead by the time they scared off the tiger and got to him – a timely reminder that nature is both beautiful and deadly.

Nothing could excuse his elder son's behaviour; not content with getting point-blank shots of the dogs feeding, he drove his elephant at the animals in order to get even closer views. Eventually, of course, they deserted their half-eaten meal and it was left to the opportunistic Vultures. The greed of this ranger – whose job it was to protect the wildlife in the park – resulted in some hungry cubs that day.

I had met most of the residents from Bandhavgarh here again in Kanha, and the two large English ladies waded into any official within ear shot at the park gates to complain, but they rapidly closed ranks to protect their own; *no*, the head ranger was not around, *yes*, there was a complaint book in the centre, *no*, it wasn't open and *no*, no one knew where the key was. It seemed that not just the tigers were betrayed in India's national parks by those responsible for caring for them.

We eventually did find tigers in Kanha, the best of which was a superb young male (my first) which looked on bored from its ledge in the bush as I got to within a few yards on my elephant.

Mind you, this encounter was nearly bettered by another male we stalked through the tracks in the woods, and which finally emerged into the open and walked nonchalantly between our jeeps. Richard and I were almost wetting ourselves with excitement at this point.

The park was also famous for its Indian bison, the world's largest species of cattle, which you were apt to bump into early in the morning as you rounded a corner in the deepest forest. You certainly wouldn't want to bump into these brooding hulks in the dark that's for sure.

The camp was better than the park for birds, being so near a water source. Whilst others relaxed in the afternoons, I was out picking off Tickell's Flycatcher, Golden Oriole, Black Naped Monarch and a tiny Purple Sunbird on its thimble-sized nest in a bush just by the restaurant in the camp. Mr Eric very proudly told me that he had found this, and that it was a first for this part of India.

The food had improved at least in Kanha, but not so the weather; after six weeks of perfect weather since leaving Sydney for the last time, one bird walk was hastily abandoned as the clouds appeared from nowhere, the wind suddenly got up and the heavens opened. In minutes the camp was awash, and one of the most vivid memories I have of Kanha is sitting in the shelter of the camp, watching the river rise before our eyes, whilst in the distance a line of classic cars made their way through the deluge and across the bridge over the river, on a rally which somehow had traversed the roads from Bharatpur to Kanha. It was one of those times when you wished you had the talent to paint, especially when the weather cleared as quickly as it had arrived, leaving long smoky trails of mist between the hills and suspended over the meadows of Kanha.

It was only a matter of days now until I was due to meet Naaz in Cape Town, and my excitement was growing daily. I was fairly sure by now that she would come, although at the back of my mind was the thought that it might only take one bold ultimatum from JJ to change her mind.

To make things worse, the phone lines never seemed to work in any of the camps I stayed in, and it had been almost a week since I had had the chance to nurture our relationship, and give her the courage to get on that plane despite her partner's obvious disapproval. It was therefore with some relief that Mr Desmond came knocking on my door after dinner one evening, saying that Naaz was on the phone in Mr Eric's office.

What I was not to know was that Mr Eric's office was also Mr Eric's bungalow, and I somehow managed to have an intimate conversation with Naaz, in the middle of nowhere, in the middle of the night, in the middle of another thunderstorm, with a pyjama-clad Mr Eric pretending he really didn't mind and pacing up and down behind me.

After that it was all down to her. I didn't know if I would get a chance to talk to her again before I left for Africa, and prayed to God that her nerve would hold.

The journey into the setting sun from Kanha to the airport at Nagpur was largely uneventful, and I began to get depressed that I was leaving the best part of India – its countryside and wildlife – behind to exchange it for the worst part – its cities and their people. It was well past midnight when I landed back at Mumbai, and the dubious comfort of the same palatial hotel I had first set foot in the night I arrived in India some three weeks before – it seemed like half a lifetime ago.

This mood was not improved by a meeting with a representative from the ground operators the next day. I had already negotiated some compensation from them because, as I had feared, not all arrangements had gone according to my itinerary. I had also agreed to take this in US dollars, since I had no more need of rupees, and was about to leave the country. This bureaucratic little jobsworth had the temerity to try and swindle me on the exchange rates, and on the compensation his company had agreed to pay me for screwing up in the first place!

It was this attitude that made dealing with officialdom, especially in the cities, in India such a hassle. If it had not been for the bewitching and bewildering tableaux as you travelled through the countryside, and the rich tapestry of its wildlife, I would have been happy to never have set foot in India again.

My final day in this conundrum of a country was to be spent in the commercial centre of Mumbai. It soon became evident that this city was very different to, say, Delhi or Agra; billed as 'Bollywood', India's answer to Los Angeles, it lived up to the name in its appearance.

The vast urban sprawl of the peninsula on which the city stood was dotted with skyscrapers and criss-crossed with freeways, each of them lined with huge billboards promoting anything and everything – it even had too much traffic and smog, just like its US counterpart. It was by far the most Western of any place I'd been in India and, in the context of its cities at least, I'm ashamed to admit that it was all the better for it. However, the contrast between the poverty in the surrounding countryside and the opulence here was never starker.

I took a cab down to the Gate of India, an impressive red sandstone Victorian monolith commemorating the visit of royalty nearly a hundred years before, and ran the usual gauntlet of insistent hawkers and beggars before boarding the boat for Elephant Island. This turned out to be a fairly average collection of caves and statues from ancient times, but it did allow me the rare opportunity at least to tick some seabirds amongst the dock complexes and huge liners – both Brown and Black Headed Gulls, Yellow Legged Gull and Gull Billed Tern.

And, as sod's law would have it, I got talking to an attractive English girl who was bored and had time on her hands, just at the very moment I began to run out of both time and rupees. Linda, a croupier on one of the liners, and I managed a quick drink together in the bar in one of the very plush hotels along the waterfront, before I had to dash back to my hotel to pack for that evening's flight.

My thoughts, as I waited out the interminable delays for my flight across the Indian Ocean to South Africa, were in turmoil. I had loved most of my time in India, and it had provided some of the highest of highlights of the entire trip. I was sure that the tigers of Bandhavgarh and Kanha would never be beaten by any other wildlife spectacular and the world outside my car window was a constant series of wonderful, colourful, thought-provoking surprises.

India had also rescued my quest for 1000 bird species from disaster. The total of 213 new birds was only slightly lower than in

Ecuador and with the running total now only a few dozen short of the magic number, I felt it was only a matter of time before I scored that one thousandth bird in Africa – if I could tear myself away from Naaz, that was.

However, India had also given me some of the lowest points on the Trip. The endemic corruption, surly attitude to Westerners and lack of regard for their own natural riches made me despise most of the Indians I'd met, and I certainly failed to form friendships with them as I had done with people in other countries on my way around the world. The fact that I was involved with an Indian girl, and thought I knew something about the culture, made this all the more surprising.

In many ways there were parallels between Naaz and the country of her forefathers which would have greatly surprised her; beautiful and complex, passionate and colourful, full of vibrant energy, often kind, sometimes cruel, but always surprising, and by turns seductive and alluring. Both were beset by contrasting moods, were sometimes intimidating and infuriating, and yet were always delightfully mysterious – definitely different to everything I was used to. A puzzle in both cases and one I felt I was never going to solve in either of them.

I flew then with Naazlin on my mind, that and the realisation that South Africa was to be my final destination on my trip of a lifetime. The only question that remained was, would she turn up?

BIRD COUNT: 955

13 An Antelope in the Shower: South Africa, March – April

I landed in Cape Town with a thousand things on my mind. For starters, this was the last country, the last adventure, and the last birds of my six-month trip. It seemed as if I had just got into the swing of it all, and now suddenly it was almost over.

Returning home, facing all my old problems again, after running away from them for all this time was also going to be difficult – after the pleasure, the pain. My mood was not helped by my first images of Cape Town; far from the exotic and vibrant African city I had expected, it was enveloped in low cloud, even obscuring Table Mountain, its most famous landmark that overlooked the city like some watchful big brother. After all the constant sunshine in Asia, this was a bit of a rude shock. And it didn't seem African at all – most of the population seemed to be white or Indian.

I needn't have worried. I should have learnt by now that first impressions of new countries are always mistaken, usually the result of the stark contrast from the place you have just left and overshadowed by the stress of travelling and settling in. By that first evening, I was revelling in the fact that here was a country where the natives were friendly, where I didn't need to worry about the food, and where I could be assured of lots of decent red wine. It reminded me a little of Sydney, but without the sunshine.

That first night was spent out in the tourist enclave of Camp's Bay, a smart suburb by the ocean where just about every white tourist congregates to take advantage of the great local restaurants and escape the crime in the tattier parts of Cape Town. Or, that was the idea at least.

Having settled in I set about putting the final parts of Operation Naaz into play. A few quick calls and a sweep of a credit card had provided me with champagne and flowers for the room the next morning, and a big white limousine for the airport.

I had decided on the flight over from India that I would contact Tim, my brother and guardian of my finances at home, to send me enough money so that Naaz and I could have the two-week holiday of a lifetime. It might be my last chance, as in a few weeks she could be living in Switzerland. We certainly were never going to get this amount of quality time together again. I wanted to give her an experience with me she would never forget.

I was nearly as nervous the following morning whilst I got ready, as I had been in London earlier that year. I thought of how high the stakes were on the drive to the airport – what if Naaz had bottled it at the last moment? Not only would it leave me at a bit of a loose end in South Africa, but the Trip as a whole would feel like a failure if the final adventure didn't come off. I steeled myself for the worst.

I watched her plane land on a far runway and taxi to the gate, and the coaches pick up the passengers and drive right past me to deliver them at immigration. I didn't see a familiar smiling brown face at the window, and the butterflies in my chest had now turned into flocks of seagulls. To make things worse, this was my first time meeting someone off an aeroplane, whereas everybody else in the crowd around the exit seemed relaxed and a dab hand at the waiting game.

I was busy wondering how to handle that first moment – play it cool? Burst into tears? Kiss her madly? Shake her hand, for God's sake? For your average British male, unaccustomed to a show of affection in public, the stress was immense, trying to find the right way to express yourself, and knowing a slip could be costly.

I tried to stay calm by imagining how many people would be ahead of her in the queue, but that number had soon been exceeded, and still no Naaz. The dark seeds of doubt were beginning to grow in my brain when, quite suddenly, there she was – grinning broadly, chatting to a cluster of people she had met on the flight and, just as typically, squeezed into a tiny top and skirt that would be revealing in Cape Town in summer, never mind London in winter.

At this point I needn't have worried about how to handle the situation; I just went into autopilot as my emotions took over. I found myself unexpectedly, and without a hint of embarrassment, rushing over to her and giving her the biggest hug and kiss I've ever given anyone. She looked great, smelt wonderful and with that first embrace all the pain and loneliness of our three months apart dissipated almost immediately.

Our driver took the long way back to Camp's Bay so that we could continue our clinch in the gigantic back seat. I couldn't quite believe she was there, and had to keep touching her just to reassure myself that she was. We sat there laughing and grinning inanely at each other all the way back to the hotel.

After the excitement had dipped below crescendo level later that day, it was time to plan our two weeks together over a lazy lunch and a bottle of wine down at the Waterfront, Cape Town's' answer to London's Covent Garden. Exploring one of the most romantic cities in the world was obviously a must, but then we intended to rent a car and spend some time in the wine region to the north of the Cape. After that, a quick visit to the far north to stay at the ludicrously impressive Palace Hotel in Sun City and, finally, an animal extravaganza of some kind to round off the tour nicely – and, of course, to ensure that I spent some time looking for that elusive one thousandth bird, rather than just down the front of Naaz's cleavage. A tough choice, I thought as I gazed across the table at her.

I had decided to go to Cape Town as opposed to other parts of South Africa because I had heard horrific tales of car-jacking and mugging in Johannesburg and the like. Cape Town was meant to be safe. I had booked into a big hotel just to be on the safe side; but how wrong can you be?

We awoke the following morning to discover that whilst we had been at the Waterfront the previous afternoon someone had broken into the room and stolen my camera. When I found out that it was missing from my rucksack, my initial thought was 'Oh God, please let the photos of India be safe!'

Fortunately the thief had been more interested in what he could sell than my undeveloped films, although I did lose a dozen or so from Mumbai that were still in the camera and, annoyingly, also the pen that Naaz had bought me and which I had used to write to her so many times over the last few months. How very spooky that it should disappear on the day that we were finally back together again, as if to say that we don't need to write anymore – now we had the real thing.

The hotel, anxious to not attract a reputation as a haven for pilfering, were unwilling to act even though they admitted it must have been one of their staff, and I let it drop thinking that I didn't want to waste any of our time together and that the holiday insurance would cover the cost of replacements. It was only after I got back I discovered the small print, which meant I was not getting any money back at all. Travellers beware!

I refused to let the episode get in the way of our first day together, despite the galling thought that I had managed to sleep on board a small boat in the Galapagos, on the floor in a house in Thailand and on a train in Vietnam without getting anything stolen, only to come here and get robbed in a palatial hotel in Cape Town.

We went for some retail therapy, bought a replacement camera and, after a couple of hours with a local travel agent, had all the rest of the trip sorted and paid for. We had gone for the very best available and paid accordingly, but I thought it was a sound investment.

This left us free to drive up to Table Mountain, now thankfully standing in all its sunlit grandeur above the city. The escarpment it is part of stretches for mile upon mile along the coast at this point. Having taken the cable car to the top we were able to see all of the city and its surrounding bays, hills and mountains in one perfect panorama in bright sunlight, framed by the deepest of blue seas and invigorated by the crisp autumn air up on the summit. For autumn it was once again, and I had spent so long and so far travelling that I had caught up with the same season I had started with six months before, on the other side of the world.

Far below you could just make out Robbin Island, where Nelson Mandela had spent so many years in captivity, and we spent a happy and romantic couple of hours exploring the rock formations on top, which at this range looked anything but as flat as a table. Periodically we would be damply enveloped in a small cloud that would suddenly appear and then disappear over the lip of the mountainside, and it seemed to us as though we were almost part of the sky.

We had lunch looking down at the city laid out below, periodically fending off the rock hyraxes that saw tourist meals in the same way the local population saw tourist bags – rich pickings. These vicious little beasts, called Cape rabbits locally but in truth more like a hamster on steroids, waited until you were poised between mouthfuls before launching a surprise attack from under the table for the plate above, and only a sharp and valiant thrust with a fork was enough to send them packing.

A journey back down to the coast brought up the first real South African birds: Rock Pigeons, Mountain Chat and gloriously confiding Red Winged Starlings on the mountain, Blacksmith Plovers on the city roundabouts, and Hartlaub's Gull and Cape Cormorant in the bay. I was hoping that just being in Africa – my fourth continent of the Trip – would bring enough new species to meet my target, as I knew it was going to be hard to tear myself away to do any proper birding over the coming weeks.

My best chance to tick some of the dozens of Cape endemic birds came the following day, when we drove a couple of hours south along the peninsula to the Cape of Good Hope. Placed on the end of its long spit of land jutting into the Atlantic Ocean, its unusual geographic location gave the Cape its own unique natural history – and birdlife.

Our journey down the Cape was interspersed with stunning views of steep cliffs and crashing seas, and once by an unwelcome wrong turn into a forbidding local black township. The mountains in this part of the world, often covered in cloud and subject to notoriously wet and wild weather patterns, were rugged yet always covered in a lush green coat of vegetation, a landscape like no other I'd seen on my travels.

The Cape itself was predictably touristy; the most obvious and imposing promontory was in fact Cape Point, its more famous cousin the Cape of Good Hope lying rather more modestly a couple of miles north-west of here. The other fallacy was that here you could see the confluence of the Atlantic and Indian Oceans. Certainly out at sea there were two distinctly different seascapes, but this had more to do with local currents and undersea rock formations than anything else. The real border between the oceans lay miles off to the east and out of sight.

Trekking up and down the windswept cliff paths revealed some of the birds I'd come to see: Cape Wagtail, Cape Sparrow, Cape Bulbul, Cape Bunting and the linguistically challenged Bokmakierie, characteristically feeding on the yellow flowers along the cliff tops.

Lower down, actually on the Cape of Good Hope, we spotted a family of Ostriches feeding amongst the dunes by the roadside, and crossing the road in front of us as if we were not there. However, our closest encounter, with a beast that made the rock hyrax look like a pussycat, came as we sitting in a café eating lunch.

As we munched our sandwiches, we were aware of several of the local park rangers shouting and throwing stones a couple of hundred yards away, chasing a group of baboons deep into the bush, their raucous cackles seeming to mock their pursuers. We thought no more of it until Naaz was wrist deep in a packet of crisps. Suddenly, a huge male baboon vaulted the wall around the café, jumped clean over the table next to us, sat next to Naaz momentarily and in a blink of an eyelid had snatched the entire packet and headed for sanctuary on the other side of the wall. This Mike Tyson of the monkey world was back in the bush and scoffing the lot before you could say 'bright red arse'. We both sat there stunned, scarcely believing what we'd just seen – a simian mugging, and so soon after our hotel burglary. Who said Cape Town was safe?

A late afternoon swim in a chilly seawater pool cooled us down, and as the sun sank lower over the ocean we were able to find time to visit the colony of Jackass Penguins at Boulder's Beach before it closed.

This was one of those wildlife extravaganzas that takes place in the midst of human habitation. Despite the fact that the dunes above their beach were now covered by swish holiday homes for rich Cape Towners, hundreds of dapper pied Penguins occupied the breeding beaches, swimming in the surf, drying in the sun or snoozing in the breeding burrows. It was a sobering thought that these were very close relatives to the Megellanic Penguins I'd seen on the other side of the Atlantic, at Peninsula Valdes in Argentina, nearly four months and over 15,000 miles ago.

If you could ignore the smell and concentrate on the spectacle, the day-to-day comings and goings in the colony were better than any soap opera, and even Naaz was amused enough to stay with me on the beach until the sun was balanced on the horizon and the temperature started to drop. This was the signal for a scrumptious seafood dinner and a chance to make plans for the trip to the wine region the following morning.

The drive north was made easy by the fantastic roads in this most Western of African countries, and with Naazlin's madcap driving at the wheel we had soon left the city sprawl behind and were driving into lush countryside, once again framed by the tall green mountains of the Cape region.

Our first stop was in Franshoeck, one of those Afrikaans names that simply cries out to be rolled across the tongue. The town itself was little more than a village, with a handful of hotels and restaurants and a small white church framed against the mountains and impossibly blue sky.

This was one of those South African quandaries – in a country with its fair share of poverty and strife, here was an oasis of calm and wealth, where only the servants were black. We were to come across several of these, and it was evident that although the political power now lay with the blacks, there were still effectively two countries inside South Africa.

Our hotel in Franshoeck was simply stunning. Built like a split-level beach house overlooking the pool, it was all vibrant bright blues and yellows, with an enormous feather bed dominating the room. The table had home-made savouries and cordials, the bathroom had champagne, candles and chocolates. It was the epitome of intimacy.

We spent a lazy afternoon briefly tasting some of the wines at a local vineyard (just so we could say we had), before polishing off a whole bottle of local Pinotage and dozing around the pool. This was luxury, and luxury I felt we both deserved.

However, amidst all of this Naaz was tortured by feelings of guilt. Although we all knew that she would return to live with her boyfriend, and she was in touch with him by phone, it was hard for her to relax knowing that he now wished he had given her an ultimatum not to come. It made me all the more determined to make this trip worth the sacrifice she had made.

The neighbouring town of Stellenbosch was our next stop. This was a far more substantial town, and the main one in the Cape wine region. It had a laid-back, worldly atmosphere, courtesy of its large student population. This was unusual in a country where many of the 'white' towns seemed provincial and rather austere, as if still living in another age, and at odds with the natural splendours of their wonderful country.

Once again the boat was well and truly pushed out, as we decamped to a large country hotel just outside the town, all sweeping lawns and wooden staircases. Always difficult to cater for with such a strict vegetarian regime, Naaz was finding eating difficult in a country whose cuisine seemed to be based around red meat, and lots of it. However, this was soon settled when the chef invited us both into the kitchen, even in the midst of preparations for a large wedding, so she could choose the ingredients from the pantry for the risotto he was to cook especially for her.

The highlight for her, though, was the visit of a local masseur later that afternoon, after a shower of rain had driven us inside. He turned up at our room, led by a member of staff – he was totally blind, and after losing his sight had learnt his art having discovered that his sense of touch was now much more accentuated. It was incredible to watch, as he moved around her prostrate body, using his hands to stroke, push or caress, moulding her flesh as he gazed sightlessly into the distance, barely uttering a word. It must have been incredible to feel his touch, and Naaz was floating on cloud nine for the rest of the day.

We drove back to Cape Town before dawn the next day, and caught our flight further north to the airport at Sun City. On the flight up I totted up the Bird List. In the wine region I had picked up suburban birds like Cape Turtle Dove, Cape White Eye and Olive Thrush, and together with those in the Cape itself, I now discovered that the total was 997! So, I thought, this was it – the one thousandth bird was to arrive on day number 159, in Sun City, South Africa. Surely that afternoon was to provide the culmination of so many months birding, and with only a week to spare.

Sun City was built in the 1970s as a haven for white tourists, and being inside a tribal homeland, was immune from apartheid's strict moral code. It developed a reputation for being an African Las Vegas, with a large casino, lavish shows and numerous fleshpots for the wide-eyed and deprived South African holiday-maker.

Since the political revolution, however, this moral code had been swept away, and South Africa is now one of the most liberal of all African countries (take one look at the number of fleshpots in Cape Town if you need evidence, but beware the countries' spiralling AIDS infection rate). Sun City today therefore was no more than one large, and expensive, tourist complex.

We were here for two reasons. First, I had come two years before, on a freebie with several other media bods courtesy of Channel 4 in London. We had dinner one night in the most expensive hotel in the resort, the Palace, and it was so ridiculously over the top that I promised myself that one day I would stay here with someone I really cared about. Naaz had given me my chance.

The other good thing about Sun City was that it had a game park right next door. Pilansberg had been set up some fifteen years before, when the wildlife service bought up local farmland and set about restocking it with animals that had once been indigenous to the region, and returned the area to its natural state of veldt and rolling hills. Now it had large self-sustaining herds of most of the large African mammels, and the hotel ran two game drives a day – my first chance to see some Big Game in Africa on this trip.

Naaz could not believe the opulence as we drove up to the Palace. Outside was an enormous fountain, filled with a golden sculpture of a cheetah attacking a herd of impala. The architecture itself was done in the style of the lost city of Zimbabwe, with towers and domes at every corner and horns and zebra hide decorating every wall. In the atrium, you looked down on the dining room, dominated by another fountain, this one filled with a herd of elephants. The square in the middle of the hotel contained a massive statue of a charging bull elephant, twice my size. Over the top it may be, but it was all too well done to be tacky. And it was a hotel that you certainly could never forget.

We spent the remainder of the morning exploring the hotel, and the man-made beach (complete with wave machine) that lay below it. By this stage my nerves were stretched by the pursuit of bird number 1000.

Number 998 was ticked off easily enough – beneath the huge arches and turrets of the hotel, there were hundreds of nesting pairs of White Rumped Swifts. Around the hotel there were several Cape Glossy Starlings – bird number 999. The next one was the big one!

It was a frustrating afternoon, having lunch, swimming, shopping, and all without a new bird in sight. Naaz seemed oblivious to the fact that at any moment my quest could be at an end.

In the end, of course, it was all an anti-climax. We had deserted the pool in favour of an afternoon nap after our early start, and were walking back up to the hotel. On my first visit to the Palace, the hotel had been surrounded by a series of pools and waterfalls, but many of these had now been drained for renovation, leaving a few sludgy concrete basins to spoil the view. This meant that a couple of species of duck I had been expecting were nowhere to be seen, but as we walked past one of the sludge pits I did catch a movement out of the corner of my eye. I looked around and there it was – *bird number 1000!*

In my mind's eye I had envisaged the one-thousandth bird to be a magnificent bird of prey perhaps, soaring over the mountains, or a multi-coloured exotic deep in the rainforest, perched in a sunlit glade. Instead I had an Egyptian Goose, looking rather sorry for itself, on a concrete island in a sludgy pool. Not only an inauspicious location, but also a bird I could have found on an estuary in Norfolk as easily as here in Africa.

It was only later that the elation crept in. 'I've done it! I've done it!' I thought – not a mind-blowing total in world birding terms, but nonetheless *my* total, and one that I had spent months compiling, in twelve countries across four continents. It was a very proud moment for me, and it was only then that Naaz could see what it really meant to me to have achieved it.

I still needed some more birds to add to the total, as I was sure that when I edited the List back home I would lose some species. There are lots that are the same species zoologically speaking, but are called different names in different countries, and I couldn't count them twice. I would need to have anther thirty or forty on the List, I felt, to still have 1000 after the edit. Could you imagine coming home thinking you'd cracked the 1000 total, and then being denied because the Great Cormorant is called all sorts of other names but can only be counted once on the List?

Fortunately our couple of trips to the game reserve at Pilansberg provided ample wildlife viewing of all kinds. Picked up outside your palatial hotel, you were transferred by jeep into the park and suddenly it was like there was no Sun City around the corner – you were deep in the African bush.

Many of the animals, as in the whole of Africa, were food sources for the predators – the ubiquitous impala, (referred to by our guide as the 'McDonald's of the veldt') plus kudu, waterbuck, red hartebeest, wildebeest and zebra – the variety was staggering. In the reservoir there were a couple of slumbering hippos, whilst every so often a giraffe appeared over the top of the thorn bush, and gazed at the jeep as if *he* was photographing *us*. But, there was a distinct lack of the predators themselves.

We did see one elephant, framed photogenically against the hills illuminated by the setting sun, and also a family group of white rhino, although minus a couple of siblings who had been killed in the most unsavoury way imaginable by being attacked by three juvenile elephants in a pubescent rage full of rampant sexual energy. I had been alone often enough on this trip to understand that kind of frustration.

Another fifteen or so birds were welcome additions here as well. The pick of these were the pair of Hamerkop gliding over the tree line, the Hadeba Ibis roosting on the reservoir islands, the Grey Louries calling mournfully from the trees around the reserve centre and the Crested Francolins that scurried along in front of the jeep, seemingly about to be crushed under the wheels at any second.

A final day soaking up the luxury around the pool, feeling for all the world as if we were a honeymoon couple (at least until her boyfriend called our room), and then we were off again. This time it was a long car journey south to Johannesburg, and the very last big adventure of our trip – the Kruger National Park, and our camp at Ngala.

The Kruger was one of the first national parks in the world, and occupies a huge swathe of bush along South Africa's eastern borders with Mozambique. Until recently most tourists had to travel into the park from outside, as it was fenced off from the surrounding countryside, and therefore there was a throng of wildlife camps and lodges of various descriptions around the perimeter.

However, in the last few years there has been a concerted effort to remove these fences, not just around the Kruger itself, but also in some places along the borders as well with Mozambique and Zimbabwe. The idea is to restore the historic migration routes for the animals, which human intrusion had taken away. There have been horrific stories as a result of this, such as herds of wildebeest throwing themselves at barbed-wire fences in an effort to fulfil their primal urges and migrate as their ancestors had done.

Thus an area the size of Wales was now more open to the wildlife tourist, and no doubt there will be more development because of it. Ngala was the first private camp to be built actually inside the park itself, and the tented camp we had booked ourselves into was brand new – we had the honour to be one of the first visitors.

To get there we had a four-hour drive to Johannesburg, a flight out to the tiny airfield at Skukuza, and then a light aeroplane from there, stopping at all the airstrips en route before finally coming to a bumpy and dusty halt at Ngala's own.

This was another world. Already you had the feeling that you were effectively cut off from the world outside, but having entered the tented camp, it was soon apparent that this did not mean that we were cut of from its amenities – far from it.

The camp was situated along a dried up riverbed, the water diverted by local towns long ago. The accommodation were tents in name only – a dozen beautifully appointed cabins, with canvas sides and roofs, complete with wooden floors, sunken baths, an enormous bed and a shower outside on the rear decking and, in our case, complete with a startled waterbuck, living up to its name, and drinking the water dripping from the shower head! It was a mini-palace and already we could tell that these last few days in South Africa were going to be an experience to be cherished for ever.

We were the only guests in a camp with twenty-two staff. We

were introduced to numerous cooks, valets, guards and waiters, and just had time to freshen up before that evening's game drive (Naaz's purple bikini, originally bought by me in Sydney, was never going to be appropriate for chasing wildlife through the bush).

We soon set off in our monster of a 4 x 4 vehicle, its 8-litre engine ready for anything the bush could throw at it. In fact, our driver Derek, and his spotter Eric, riding shotgun precariously on a seat out in front of the bonnet, had no compulsions when it came to going off road, mowing down bushes and small trees in pursuit of animals. I wondered how long the environment could stand this kind of treatment.

Meanwhile the back of the vehicle was raised, so that we could see everything from a lofty viewing point. I was particularly keen to see a leopard, this being the only one of the Big Five – lion, leopard, rhino, elephant, and water buffalo – I had never seen. Derek was in radio contact with another couple of rangers in vehicles like ours, and within minutes we had come across two juvenile lions dozing in the long grass as the sun began to set.

Soon after this, Derek really got the adrenaline running when he announced that one of the rangers had discovered a young leopard a couple of miles away. It was almost dark as we sped along the dirt tracks and crashed through the undergrowth, with Naaz and I having to bob and weave in our seats to avoid low hanging branches, and the large nocturnal insects that threatened serious damage in a collision.

As we got nearer, Eric deserted his seat at the front and took refuge next to Derek, and we slowed to a halt when we saw a ranger ahead – on foot no less – with a large flashlight. He waved us on, and we crept forward as stealthily as you can in a 5-ton monster of a vehicle. Then Derek saw it in the beam of his powerful flashlight, a half-grown leopard cub, more confiding then we could ever have thought, letting us get to within 10 yards before ambling off into cover, calling plaintively for his mother nearby.

It was a magical encounter, made even more special as we drove home by some of the nocturnal wildlife on show – a black-backed Jackal on a hunting foray, Fiery Necked Nightjars hawking in our headlights like giant moths, a Pearl Spotted Owl standing vigilant on a tree over the road. It was a wonderland and so beautiful in its profusion of natural riches that it took your breath away.

Dinner that night was for just the two of us, around a roaring open fire on the riverbank, with the small swimming pool and its waterfall tinkling away behind us. We felt sorry that the staff had to go to all this trouble just for us, but at these prices I guess nothing was too much trouble. We had drinks before dinner whilst our order was taken – even here in the bush there was a choice of dishes – and then sat back and were waited on by waiters in bow ties and white dinner jackets. Dinner itself was only interrupted by one of the many camp guards taking Naaz and I aside to show us a juvenile hippo that came to feed regularly on the camp lawns – for all the world like a big grey cow, munching away in the darkness and surprisingly agile out of the water.

The hippo kills more people in Africa every year than any predator, and it proved that the guards around the camp were not just a precaution against *human* infiltration after dark.

That night I lay awake in our big bed, watching the canvas roof billow in the breeze and the moonlight coming through the tent walls. Outside I could hear the throaty roar of several lions, and closer, the maniacal hysterics of a pack of hyenas, sounding like some spirits in the night. I looked across and watched Naaz sleep, her raven hair framing her noble face, her big brown eyes closed now but still beautiful under those impossibly long lashes. As my journal says, 'does it get any better than this?' Certainly I had never been happier in my life.

The next day was a special one for Naaz and I – exactly a year since our first sexy lunch in London together, since we both sensed that special chemistry together. It didn't seem possible that I had spent half the time since travelling the world, and yet here we were on safari together in the Kruger a year later.

We were awoken before dawn by one of the guards, leaving lanterns and fruit on our veranda and then escorting us through the darkness to the main building where meals were served. This was like a large den, with a deck out over the riverbed and a large canopy shielding it all. A quick cup of coffee as the sun rose and we were on our way again.

It was an eerie experience driving through the half-light, watching the shapes of animals slowly coming into view as the dawn broke over the bush. This time I tried to focus on the bird life, which was harder to find than the animals in such a vast area. Nevertheless, classic safari species were soon added to my List, like the Red Billed Oxpeckers on a giraffe, the Paradise Wydahs dragging their enormous tails across the grasslands, and the Spotted Dickop and Secretary Birds stalking for insects amongst the herds of impala and zebra.

In truth though, the best birds were to be found around the camp; the main building played host to a noisy pair of Red Billed Wood Hoopoes, whilst along the riverbank one could find Striped Cuckoos, Woodland Kingfishers and Long Tailed Shrikes. But perhaps the best birds seen from here were over the far trees one lunchtime – two enormous Cape Griffon Vultures, playing on the thermals.

After the morning drive, it would be back to the camp for a very hearty breakfast, before a doze and a swim, lunch and then another drive – the perfect mixture of wildlife-watching and relaxation. We were joined that night by a honeymoon couple, and did our best to reassure the rather nervous new bride that she was not going to get bitten by a tsetse fly or eaten by a lion. The evening ended in even more dramatic form than it had begun, with a tremendous thunderstorm ensuring little sleep for anyone under canvas that night.

Our new travelling companions certainly provided as with good luck the following morning. One of the things that ever person on safari wants to see during their trip is the Big Five, and in the Kruger this is usually guaranteed if you spend more than a couple of days searching. Naaz and I still needed a water buffalo and a rhino to complete our own set at Ngala.

We spotted the female leopard and her half-grown cub early that morning, and spent a rather unsavoury hour following them across ravines and inside bushes at a range I felt was far too intrusive. We then encountered the same two lions we had seen on our first drive, followed quickly by a typically aggressive and bad-tempered bull Elephant crashing around in the undergrowth to show us who was boss.

We now had three of the five in less than two hours searching, and we all decided that if we could get all five in a single drive, then it was an experience that would not easily be forgotten. Nevertheless it was a long wait until we found some white rhino, and sneaked up behind them (they have notoriously poor eyesight) as they fed in the deep bush.

This then just left the water buffalo, an animal I had only seen once before, in Botswana. We had not even seen tracks of one at this point, and were giving up hope of completing the set, when we crossed the dry riverbed for the last time on our way home. Suddenly Eric pointed to the right triumphantly, and sure enough about 400 yards away there was a brown lump on the sand – our water buffalo and the last of the Big Five!

We drove slowly up to it, but needn't have worried – this individual was here for the duration, seemingly dozing and digesting its breakfast. Immobile it may have been, but they all counted, and I took almost as much satisfaction from completing the Big Five in one drive as I did the 1000 birds in six months. But it was certainly a near run thing; that Buffalo was to be the only one we saw in the entire time we were in South Africa.

I felt rather sorry for our fellow guests, since they had seen all the Big Five in their very first game drive – everything after that was going to be an anti-climax. I don't think they truly understood the magnitude of what they had just experienced.

We spent a fascinating lunchtime talking with the manager of the new camp, a young white girl for whom this was her first assignment. Kim loved the bush, had a great rapport with the staff and was happy to answer our questions about her life under apartheid. Despite being white, her family was part of the protest movement and she told us stories of how her father was put under police surveillance, and the windows of her house smashed. She said that Mandela coming to power lifted a heavy yolk off the South African people, and even now she was only slowly coming to terms with how different life was without the oppressive regime in place. It really made you think how lucky we are to live in a relatively liberal and tolerant society in our own country.

Our final game drive was going to have to work hard to live up to its predecessor, and for a long time it threatened to be a damp squib. But, as usual, Derek and Eric came up trumps – in fact we came to suspect that the animals were simply models, which could be switched on and off by the rangers, so seldom did they fail us.

Seeing a circling White Backed Vulture, we crashed through the bush in the hope of finding a kill, but never dreaming that we would stumble across a lion that would be so happy for us to watch him eat. This young animal had stolen an impala from the jaws of a local leopard and, having done all the hard work, was in no mood to let anything spoil his lunch. We inched forward towards the bush he was lying beneath, until we were so close we could smell the entrails, and could have touched him had we been foolhardy enough. We could hear him crunching his way though the bones, and got some incredible camera shots even with our tiny lenses.

There was still one last momentous evening ahead, and as darkness fell we stopped the vehicle and broke open the drinks cabinet on board. What a fantastic end to the trip to the Kruger; drinking a vodka and tonic as you watch the sun set over the acacia trees – the African sky really does look so much bigger – and listening to the lions roar their goodnights in the bush around you.

I had often regretted leaving places on this Trip, but none so much as the wrench of leaving Ngala. We bade tearful farewells to all the staff one by one, and said goodbye to our palatial tented home for the last time. Of all the places on my tour, this was the one I'd most like to go back to.

A Nile crocodile and a Marabou Stork briefly enlivened our journey by road back to Skukuza as we crossed a river deep in the Kruger, but it was with a heavy heart that we flew back to Cape Town that afternoon.

My mood was not helped by the disappearance of my luggage on the way back, especially as all my gifts for home and photos from India were inside. I have only lost my luggage on three occasions, two of them being in South Africa, but at least they turned up safe and sound (and unopened) on the next flight.

Things took a turn for the better once we were back in Camp's Bay, this time opting for a funky little guest house overlooking the ocean that had been recommended to us. We both lit a cigar and watched the sun turn from yellow, to orange, to crimson and, finally, to purple as it set over the Atlantic. Only yesterday it had been setting over a pride of lions in the Kruger – what a country.

We had great plans for our final day before our return to the UK, if only to dispel my sense of foreboding that not only was the Trip all but over, but I was also handing Naaz back to her boyfriend. Although I had known this was always going to be the case, it still seemed criminal after we had been so close for so long.

In the end, a very English autumnal day and Naaz's stomach upset presented us with a strange end to proceedings; we didn't risk the boat out to Robbin Island in the windy squall, and settled for a trip around the harbour instead. It was far too cloudy for a good view of Table Mountain from the sea, but at least we did see the Cape fur seals, Heaviside's dolphin and the final bird of the Trip – a Common Tern, once again one of the most common birds back home. A muted ending to a List that had been my passion and constant companion these last twenty-four weeks, and especially odd with poor Naaz being so violently sick next to me.

Our last day frittered away as we shopped for African curios to take home and went out for our final dinner together. There was so much I wanted to say, but I did not need to. It was a rather quiet affair, with us both in reflective mood and deep in thought, but we were never more than inches apart all night.

The morning of our flight home was rather like the last day at school, or more recently, my last day in the office – disjointed, rushed, almost surreal, especially to think that the following day I would be at home in London (that is, if I had a home) and that the Trip would be finally and irrevocably *over*.

I remember sitting in stunned silence at the airport, almost lobotomised by the realisation that it had all come to an end. Naaz slept all the way home, and before I also succumbed to sleep I tried very hard to remember every single second of the trip, as if to capture it all in my internal memory for posterity.

I mulled over the statistics – twelve countries, four continents, over 1000 birds, nearly six months of my life and more money than I was prepared to admit to.

I played games then to try and make some sense of it all. Best city? Had to be Sydney. Worst city? Surely Atlantic City. Best for birds? Undoubtedly Ecuador, with India a close second. Best restaurants ? South Africa, along with the best hotels. Best bird of the Trip? Far, far too many to even begin to choose from.

I also couldn't figure out which had been the best country overall, it was simply asking too much of a hugely varied experience. I had liked them all, even loved some of them, and was already mentally compiling a list of those I would like to go back to and other ones to visit that I wish I'd had the chance to see on this Trip.

We finally emerged from the clouds over Heathrow, and I looked down at the landscape below, the first sign of spring already evident in the greening fields and woods. When I had left, the leaves had been auburn as autumn approached. Somewhere down below was my darling son, James, whom I had only spoken to half a dozen times since I left – was that his mum's fault for not letting me or mine for not trying hard enough? At least I could console myself with the fact that he had had postcards, stamps and money from every country I visited. I prayed that he would understand, and forgive me for leaving him behind whilst I fulfilled that lifetime ambition of mine.

Also down there were my Mum and Dad. Although we were not as close as some, I knew they would have been worrying about me and I was glad that I could put their minds at rest by returning home in one piece.

Finally, there was my life down there; everything I'd ever known, my career, my friends, relatives, hobbies and, yes, hopes and fears for the future.

I felt the wheels hit the Heathrow tarmac and heard the screech of the tyres. Naaz stirred beside me. What now, I thought. What on earth do I do now?

FINAL BIRD COUNT: 1032

14 Home, April

As our plane taxied to the gate, I tried hard to swallow all these anxieties. Naaz asked me if I was looking forward to seeing my family again, and I was conscious that I had overlooked the human angle of my homecoming amidst all the disappointment of the Trip being over, and my worries about the future.

I should have known better. Naaz had arranged for my parents to meet me unexpectedly at Arrivals, and I have to admit that seeing my Mum's tearful and smiling face was a real joy as she embarrassingly waved a big 'Welcome Home' banner at me across a crowded concourse. My Dad, much more a bloke, was a lot more reserved with a simple 'Hello, son' as we let the women do the emotion thing – after all, we *are* in England and this *is* in public!

The drive back to Naaz's flat was typically surreal, through the tired suburbs of West London, when I was so used to looking out of a car window and seeing a fresh new experience at every turn.

Naaz made tea and I quickly ran through the highlights of the last few months, and began to realise that whatever I did I could never recapture the vivid adventures of my travels, even with the 850 photographs I had taken to help me. It was weird having Naaz sat on my lap as I spoke to my parents, and we were both aware that although this seemed natural to us, this was now forbidden territory back here in the UK. It was at that moment I think I handed her back to JJ in Switzerland.

My parents sportingly left us to say our farewells in private. Leaving Naaz behind was amongst the hardest things I ever had to do, and as I did so a massive sadness broke over me -- I suddenly knew that things would never be this good between us again.

I arrived back at my brother's house, where I was to bed down for a while, and tried to take everyone through the photos in a whistle-stop tour of my adventures, but I felt impotent and infuriated knowing that I could not begin to do it justice. My photos were very personal to me and each one shone with some bright memory from a far away place, but to everyone else were just another set of holiday snaps.

The day got stranger, as that afternoon Tim and his wife, Jo, and I went to his local bar and I ate my first British Sunday lunch for ages, whilst watching my team, Spurs, getting humbled by Arsenal, their traditional rivals. Under normal circumstances I would have been on my feet screaming encouragement to my team but, instead, I wanted to stand up and scream at everyone 'Hey, I'm back, can I tell you what a fantastic place the world is, what a great time I've had? Can I tell you about all the people I met, the animals and birds I saw?' The truth was, it was only me that was really interested.

The next few days were a blur. I rang round my friends and, of course, spoke to James. I had to find a flat to live in and, fortunately, one fell into my lap fairly quickly. Moving in, and rediscovering all my stuff that I had put into storage months before (it felt like years), was a cathartic experience. I kept discovering bits of news, sometimes from months before, that I'd missed whilst I'd been away. I felt like some kind of hermit.

My ex-wife lived up to my expectations and made me wait before seeing James again – after all, I couldn't just swan back into the country after six months away and expect to see James at the drop of a hat now could I? When we did catch up I was suddenly aware of how resilient kids can be – he was taller, but the same son I'd left behind the previous year. He barely seemed to notice I'd been away, although he was chuffed with the gifts I'd sent him from my travels. He, too, patiently indulged me as I went through each photo, trying to impress on him how wonderful the world can be, but he was more interested in telling me how his football was going, and it was me that ended up indulging him.

I set about buying a car, my first ever, as my company had been supplying me with a car for the last twenty years. I rang around old colleagues and popped in to say 'Hi' when I was up in London. With a few exceptions, these chats were polite but banal. The truth was, as I no longer represented big clients' advertising budgets, I wasn't as big a friend to some of these people as I thought I was.

Two weeks after arriving home I was amazed and a little

depressed at how quickly I had rebooted my old life, and how comfortably I had seemingly slotted back into it – after all that, were all my travels and adventures so meaningless that it hadn't changed a thing about me? The truth was, I had only been away six months and my life was really the product of the other forty years or so. This was my reality, not chasing birds in Ecuador or watching tigers in India.

I spent hours poring over the photos, trying to squeeze every memory and last drop of meaning out of them. I carefully edited them down to a mere 500 or so, put them in albums, and had some blown up and stuck on my walls, where I can look at them. If I closed my eyes briefly I could still sometimes smell the Galapagos salt spray or hear the birdsong echo around the rainforest. It gradually allowed me to wean myself off my dream world and put me back in the real one.

I did the same with the Bird List, obsessively going through my records and my field guides, and carefully editing the names to weed out double counting and zoological anomalies. The final total of 1032 felt like a big fat worthwhile kind of number. Given that my life list stood at just over 1200 when I went, and that amongst the 1032 were 610 new ones, I felt that the return on just that basis alone was worthwhile. In six months, I had seen almost as many birds as I had in the forty previous years and managed to increase my lifetime total by 50 per cent in the process.

Apart from the desire to write, I never did emerge from the jungle with that blinding flash of inspiration I had hoped for. I was just as confused as to what the second half of my life was for as I was before I went. My other promise to myself, to learn a second language, is even now languishing in the 'things to do one day' shelf in my imagination.

Far from assuaging my wanderlust, of course, the Trip merely accentuated it. I now knew just what lay out there, that it could be experienced by the brave and adventurous, and that it wasn't just the subject of documentaries on television – it could be made real if you wanted it enough. And I'd met a lot of people, many of whom subscribed to my original idea of 'come on, there must be more to life than this!'

This made me feel more comfortable with the notion of building another trip, not this time an extravaganza like the original, but a series of destinations and experiences that were a target for me to work towards, and a dream to see me through the humdrum existence of life.

For me, this included the tropical birds and lodges in Trinidad, the unspoilt wildlife reserves in Namibia, the fantastic and unique bird life in the Himalayas in Nepal and the barren majestic wastes of Antarctica. There were also the places I could imagine returning to – beautiful Sanibel Island in Florida, wonderful La Selva Lodge in the Ecuadorian Amazon basin, Sydney of course, the awesome Southern Alps in New Zealand and, without doubt, Ngala in the Kruger. So many places, so little time.

I gave up a lot to go away on this trip – a relatively attractive career, my young son, the girl I loved and my own piece of mind. I didn't regret it at all. It had been a fantastic education in how the rest of the world lived outside my cosy enclave of media and London and England. I came back to see that little had changed, except perhaps me, richer than any millionaire in terms of experiences and memories.

But I had no job, no girl and no idea about what I should do next. So would I do it all again tomorrow if I had the chance again?

Well, what do you think!?

15 Postscript: One Year Later

Naaz and I were able to spend three more months together before she eventually emigrated to Switzerland. We used the time to underwrite our feelings for one another with some romantic evenings together, and wild weekends away in Le Touquet and Amsterdam, a castle in Sussex and Britain's oldest inn in the Cotswolds.

When she eventually did leave, I was there at the airport to see her off. As she disappeared from view it felt like I'd lost a limb. Three weeks later we were back together again, she being desperately homesick and me enjoying my first visit to Switzerland, despite a reluctant boyfriend in the background.

We were able to see each other every month, with her coming to London and me meeting her in Lausanne or Geneva, and when we were together we spent half our time trying to work out why we were apart in the first place. It was only some six months after she left that things began to finally change. Things with JJ were not good, and Naaz finally grew to understand what everyone else had known for some time – that you can only love one person at a time and expect it to work. For me, I needed to try and move away from the exhausting process of keeping up a doomed and remote long-distance relationship; this was a perfect juncture, sad though it inevitably was.

At the time of writing Naaz is giving her relationship in Switzerland a second try. We remain in love if not lovers. I am searching for a new girlfriend, perhaps one from another country and culture. We shall see.

The recession in the advertising industry was still deepening when I returned home, and I heard of only a handful of jobs during that first summer back, none of then especially suitable. In the autumn, my best friend, Colin, set up a marketing consultancy and invited me to join him. I am learning to adapt to working from home and being self-sufficient after a lifetime of working in a busy office environment and seeing others get rich on my labours.

I bought a laptop computer and began writing at the same time, and discovered a publisher through a friend almost by accident. It's taken nine months of enjoyable but hard work to get to this final page.

I lost the urge to go birding for some months after I returned from the Trip. It was only a very good autumn migration that tempted me back into the field, where I picked up thirteen fantastic new UK ticks in two frenetic months, helped by the flexibility of working as a consultant.

Last Christmas I went birding to Trinidad, the first, but by no means the last, of the destinations on my new trip list.

Alphabetic Bird List by Country

The list below is the product of my notebooks and the field guides in use at the time. I've tried to use up-to-date bird names wherever possible and to weed out any duplicate species. Please bear with me if any have slipped through the net.

Canada

American Crow
American Robin
American Wigeon
Bald Eagle
Belted Kingfisher
Black Capped Chickadee
Black Scoter
Black Turnstone
Bonaparte's Gull
Brandt's Cormorant
Bushtit
California Gull
Canada Goose
Chestnut Banded Chickadee
Common Loon
Common Murre
Dark Eyed Junco
Double Crested Cormorant
Glaucous Winged Gull
Golden Crowned Kinglet
Great Blue Heron
Harlequin Duck
Herring Gull
Horned Grebe
House Sparrow
Long Tailed Duck
Mallard
Mew Gull
North Western Crow
Northern Goshawk
Pacific Loon
Pelagic Cormorant
Pigeon Guillemot
Pileated Woodpecker
Raven
Red Breasted Merganser
Red Phalarope
Red Tailed Hawk
Red Throated Loon
Ruby Crowned Kinglet
Sanderling
Semi-Palmated Plover
Spotted Towhee
Starling
Steller's Jay
Surf Scoter
Western Grebe
Western Gull
White Winged Scoter
Winter Wren
Wood Duck

TOTAL = 51

USA

American Avocet
American Bittern
American Coot
American Green Heron
American Kestrel
American Oystercatcher
American Pipit
American White Pelican
Anhinga
Black Bellied Plover
Black Brant
Black Skimmer
Black Vulture
Black Crowned Night Heron
Blue Grosbeak
Blue Jay
Blue Winged Teal
Blue-Gray Gnatcatcher
Boat Tailed Grackle
Broad Winged Hawk
Brown Pelican
Carolina Chickadee
Carolina Wren
Caspian Tern
Cattle Egret
Cedar Waxwing
Chipping Sparrow
Collared Dove
Com. Nighthawk
Common Goldeneye
Common Grackle
Common Moorhen
Common Yellowthroat
Cooper's Hawk
Dark Eyed Junco
Downy Woodpecker
Dunlin
Eastern Bluebird
Eastern Meadowlark

Eastern Phoebe
Eastern Wood Pewee
Eurasian Teal
Feral Pigeon
Field Sparrow
Fish Crow
Florida Scrub Jay
Forster's Tern
Gadwall
Glossy Ibis
Golden Eagle
Gray Catbird
Green Winged Teal
Great Cormorant
Great Egret
Great Tailed Grackle
Greater Black Backed Gull
Greater Yellowlegs
Hooded Merganser
House Finch
Killdeer
Lesser Yellowlegs
Laughing Gull
Lesser Black Backed Gull
Little Blue Heron
Loggerhead Shrike
Long Billed Dowitcher
Magnificent Frigatebird
Merlin
Mourning Dove
Mute Swan
Northern Cardinal
Northern Flicker
Northern Gannet
Northern Harrier
Northern Mockingbird
Northern Rough Wing Swallow
Northern Shoveler
Osprey
Palm Warbler
Pectoral Sandpiper

Peregrine Falcon
Pied Billed Grebe
Pintail
Piping Plover
Prairie Warbler
Purple Gallinule
Red Bellied Woodpecker
Red Headed Woodpecker
Red Knot
Red Shouldered Hawk
Red Winged Blackbird
Reddish Egret
Ring Billed Gull
Roseate Spoonbill
Royal Tern
Ruddy Duck
Ruddy Turnstone
Sandwich Tern
Savannah Sparrow
Semi-Palmated Sandpiper
Sharp Shinned Hawk
Shiny Cowbird
Short Tailed Hawk
Snail Kite
Snowy Egret
Song Sparrow
Swainson's Hawk
Swamp Sparrow
Tree Swallow
Tri-Coloured Heron
Tufted Titmouse
Turkey Vulture
Virginia Rail
White Breasted Nuthatch
White Crowned Pigeon
Western Sandpiper
White Ibis
White Throated Sparrow
Willet
Wood Stork
Yellow Crowned Night Heron

Yellow Rumped Warbler

TOTAL = 122

BIRD COUNT = 173

Ecuador

Amazonian Umbrella Bird
Andean Gull
Andean Hillstar
Andean Lapwing
Andean Quail
Andean Spinetail
Audubon's Shearwater
Bar Winged Cinclodes
Barn Swallow
Bat Falcon
Bay Headed Tanager
Black Band Woodcreeper
Black Billed Thrush
Black Bushbird
Black Capped Antshrike
Black Capped Becard
Black Capped Donacoblus
Black Caracaras
Black Crowned Tityra
Black Faced Antbird
Black Faced Anthrush
Black Faced Dacnis
Black Fronted Nunbird
Black Spotted Barbet
Black Striped Sparrow
Black Tailed Trogon
Black Throated Mango
Black Throated Trogon
Blackpoll Warbler
Blue and Yellow Macaw
Blue Crowned Manakin

Blue Crowned Motmot
Blue Dacnis
Blue Footed Booby
Blue Gray Tanager
Blue Headed Parrot
Boat Billed Flycatcher
Buff Throat Woodcreeper
Buff Throated Foliage Gleaner
Buff Throated Saltator
Chestnut Bellied Seedeater
Chestnut Winged Foliage Gleaner
Chestnut Woodpecker
Cinerous Antshrike
Cinnamon Rumped Foliage Gleaner
Cinnamon Throated Woodcreeper
Cobalt Winged Parrotlet
Collared Plover
Common Cactus Finch
Common Noddy
Common Piping Guan
Cream Coloured Woodpecker
Crested Oropendola
Crimson Creasted Woodpecker
Dark Billed Cuckoo
Dark Rumped Petrel
Double Toothed Kite
Drab Water Tyrant
Dusky Capped Flycatcher
Dusky Headed Parrotlet
Dusky Throated Antshrike
Eared Dove
Elliot's Storm Petrel
Fasciated Antshrike
Ferruginous Antbird
Flame Creasted Tanager
Forest Elaenia
Fork Tailed Palm Swift
Fork Tailed Woodnymph
Fulvous Shrike
Galapagos Dove
Galapagos Hawk

Galapagos Martin
Galapagos Mockingbird
Galapagos Penguin
Giant Cowbird
Gold Collared Toucanet
Golden Bellied Chlorophonia
Golden Headed Manakin
Gray Coloured Flycatcher
Gray Elaenia
Gray Fronted Dove
Grayish Mourner
Great Antwren
Great Frigatebird
Great Kiskadee
Great Potoo
Great Thrush
Greater Ani
Greater Flamingo
Greater Yellow Headed Vulture
Green and Gold Tanager
Green and Rufous Kingfisher
Green Emerald Hummingbird
Hoatzin
Hood Mockingbird
Ivory Billed Aracari
Ladder Tailed Nightjar
Lafresnay's Piculet
Large Billed Tern
Large Cactus Finch
Large Ground Finch
Large Tree Finch
Lava Gull
Lava Heron
Least Sandpiper
Lemon Throated Barbet
Lesser Kiskadee
Long Billed Woodcreeper
Long Tailed Hermit
Madeiran Petrel
Many Banded Aracari
Maroon Tailed Parakeet

Masked Booby
Masked Crimson Tanager
Masked Tanager
Mealy Parrot
Mottle Backed Elaenia
Olive Spotted Hummingbird
Opal Crowned Tanager
Orange Bellied Euphonia
Orange Winged Parrot
Oriole Blackbird
Pale Vented Pigeon
Paradise Tanager
Paraque
Pink Throated Becard
Piratic Flycatcher
Plain Capped Ground Tyrant
Plain Throated Antwren
Plain Xenops
Plumbeous Antbird
Plumbeous Pigeon
Plumbeous Sierra Finch
Purple Throated Fruit Crow
Pygmy Kingfisher
Red Bellied Macaw
Red Billed Tropicbird
Red Capped Cardinal
Red Eyed Vireo
Red Footed Booby
Red Stained Woodpecker
Red Throated Caracaras
Ringed Kingfisher
Ringed Woodpecker
Rio Suno Antwren
River Tyrannulet
Ruddy Pigeon
Rufous Bellied Euphonia
Rufous Capped Anthrush
Rufous Collared Sparrow
Rufous Motmot
Rufous Tailed Xenops
Rufous Vented Chachalachas

Rusty Belted Tapaculo
Southern Rough Winged Swallow
Sand Coloured Nighthawk
Sapphire Quail Dove
Scale Backed Antbird
Scale Breasted Woodpecker
Screaming Piha
Short Billed Antwren
Short Billed Honeycreeper
Short Eared Owl
Slate Coloured Coot
Small Ground Finch
Small Tree Finch
Smooth Billed Ani
Social Flycatcher
Solitary Thrush
Sooty Antbird
Sooty Robin
Spangled Cotingha
Speckled Teal
Spix's Guan
Spix's Woodcreeper
Spot Winged Antshrike
Spotted Sandpiper
Squirrel Cuckoo
Straight Billed Hermit
Striated Heron
Striped Foliage Gleaner
Striped Manakin
Striped Woodcreeper
Swallow Tailed Gull
Swallow Tailed Kite
Swallow Tanager
Swallow Winged Puffbird
Tawny Bellied Screech Owl
Tawny Headed Swallow
Tropical Gnatcatcher
Tropical Kingbird
Vermillion Flycatcher
Violaceous Jay
Wandering Tattler

Warbler Finch
Waved Albatross
Wedge Billed Woodcreeper
Wedge Rumped Petrel
Whimbrel
White Banded Swallow
White Bearded Hermit
White Browed Purpletuft
White Cheeked Pintail
White Flanked Antwren
White Fronted Nunbird
White Necked Heron
White Necked Puffbird
White Shouldered Antbird
White Shouldered Antshrike
White Tailed Toucan
White Tailed Trogon
White Throated Kingbird
White Throated Toucan
White Vented Euphonia
White Winged Swallow
Wing Barred Manakin
Wire Tailed Manakin
Yellow Bellied Dacnis
Yellow Bellied Tanager
Yellow Billed Jacamar
Yellow Browed Sparrow
Yellow Crowned Parrot
Yellow Crowned Tyrannulet
Yellow Headed Caracaras
Yellow Legged Thrush
Yellow Ridged Toucan
Yellow Rumped Cacique
Yellow Throat Woodpecker
Yellow Warbler

TOTAL = 230
BIRD COUNT = 403

Argentina

Bananaquit
Black Necked Swan
Blackish Oystercatcher
Blue and White Swallow
Blue Eyed Cormorant
Blue Winged Parakeet
Brown Crested Martin
Brown Hooded Gull
Brown Pintail
Buff Necked Ibis
Cattle Tyrant
Chalk Browed Mockingbird
Chestnut Eared Aracari
Chilean Flamingo
Chimango Caracara
Chubut Steamer Duck
Cinnamon Teal
Coscoroba Swan
Crested Duck
Dolphin Gull
Double Collared Seedeater
Elegant Crested Tinamou
Epaulet Oriole
Field Flicker
Glittering Bellied Emerald
Golden Crowned Warbler
Gray Breasted Martin
Gray Hooded Flycatcher
Gray Hooded Sierra Finch
Gray Rumped Swift
Great Dusky Swift
Great Grebe
Green Headed Tanager
Guanay Cormorant
Guira Cuckoo
Hooded Siskin
Kelp Gull
King Cormorant
Lake Duck

Lesser Rhea
Lineated Woodpecker
Long Tailed Tyrant
Magellanic Oystercatcher
Magellanic Penguin
Magellanic Plover
Magpie Tanager
Neotropic Cormorant
Olivaceous Woodcreeper
Patagonian Mockingbird
Patagonian Yellow Finch
Picui Ground Dove
Plumbeous Kite
Plush Crested Jay
Red Billed Hawk
Red Breasted Toucan
Red Crested Cardinal
Red Crested Finch
Red Fronted Coot
Rosy Billed Pochard
Ruby Crowned Tanager
Rufous Bellied Thrush
Rufous Hornero
Scarlet Rumped Cacique
Screaming Cowbird
Sepia Capped Flycatcher
Silvery Grebe
Slaty Breasted Wood Rail
Snowy Sheathbill
South American Stilt
South American Tern
Southern Beardless Tyrannulet
Southern Giant Petrel
Southern Lapwing
Southern Martin
Southern Wigeon
Streaked Flycatcher
Surucua Trogon
Toco Toucan
Tropical Parula
Upland Goose

Versicoloured Emerald
White Banded Mockingbird
White Bearded Manakin
White Browed Blackbird
White Faced Ibis
White Shouldered Fire-Eye
White Winged Coot
Yellow Fronted Woodpecker

TOTAL = 88
BIRD COUNT = 491

Brazil

Pale Breasted Thrush
Roadside Hawk

TOTAL = 2
BIRD COUNT = 493

Australia

Australasian Wood Duck
Australian King Parrot
Australian Magpie
Australian Pelican
Australian Pipit
Australian Raven
Australian White Ibis
Bar Shouldered Dove
Barking Owl
Black Breasted Buzzard
Black Faced Cuckooshrike
Black Kite
Blue Faced Honeyeater
Blue Winged Kookaburra
Brown Cuckoo Dove
Brown Honeyeater
Brown Songlark
Brown Treecreeper

247

Combcrested Jacana
Common Mynah
Crested Pigeon
Crimson Finch
Crimson Rosella
Crimson Rosella
Dollarbird
Double Barred Finch
Dusky Honeyeater
Dusky Woodswallow
Eastern Curlew
Eastern Reef Egret
Eastern Rosella
Eastern Spinebill
Eastern Yellow Robin
Fairy Martin
Figbird
Forest Kingfisher
Fork Tailed Swift
Galah
Golden Headed Cisticola
Golden Whistler
Green Backed Gerygone
Green Pygmy Goose
Grey Butcherbird
Grey Crowned Babbler
Grey Fantail
Hardhead
Helmeted Friarbird
Hooded Robin
Jabiru Stork
Jacky Winter
Laughing Kookaburra
Lesser Crested Tern
Little Corella
Magpie Goose
Magpie Lark
Masked Lapwing
Nankeen Kestrel
New Holland Honeyeater
Noisy Friarbird

Noisy Miner
Olive Backed Oriole
Orange Footed Scrubfowl
Oriental Cuckoo
Pacific Black Duck
Partridge Pigeon
Peaceful Dove
Pheasant Coucal
Pied Currawong
Pied Imperial Pigeon
Radjah Shelduck
Rainbow Bee-Eater
Rainbow Lorikeet
Red Browed Finch
Red Rumped Parrot
Red Tailed Black Cockatoo
Red Wattlebird
Red Whiskered Bulbul
Red Winged Parrot
Rufous Banded Honeyeater
Rufous Whistler
Sacred Kingfisher
Satin Flycatcher
Silver Crowned Friarbird
Silver Gull
Southern Boobook
Spotted Turtle Dove
Straw Necked Ibis
Sulphur Crested Cockatoo
Superb Fairy Wren
Superb Lyrebird
Swamp Harrier
Tawny Frogmouth
Torresian Crow
Varied Triller
Variegated Fairy Wren
Wedge Tailed Shearwater
Welcome Swallow
Whistling Kite
White Bellied Cuckooshrike
White Bellied Sea Eagle

White Breasted Woodswallow
White Cheeked Honeyeater
White Eared Honeyeater
White Gaped Honeyeater
White Plumed Honeyeater
White Winged Chough
Willy Wagtail
Yellow Billed Thornbil
Yellow Faced Honeyeater
Yellow Oriole
Zebra Finch

TOTAL = 111
BIRD COUNT = 604

Malaysia

Asian Palm Swift
Black Naped Oriole
Black Nest Swiftlet
Blue Throated Bee-Eater
Brown Shrike
House Crow
House Swift
Java Myna
White Rumped Munia

TOTAL = 9
BIRD COUNT = 613

New Zealand

Arctic Skua
Australasian Crested Grebe
Australasian Gannet
Australasian Shoveler
Banded Dotterel
Bar Tailed Godwit
Bellbird
Black Billed Gull

Black Browed Albatross
Black Fronted Tern
Black Swan
Blackbird
Brown Creeper
Buller's Shearwater
Bullers Mollymawk
Cape Pigeon
Chaffinch
Dunnock
Fairy Penguin
Fluttering Shearwater
Goldfinch
Greenfinch
Grey Duck
Grey Warbler
Hutton's Shearwater
Kea
Little Black Shag
New Zealand Kingfisher
New Zealand Pigeon
New Zealand Robin
New Zealand Scaup
New Zealand White Capped
Mollymawk
Northern Giant Petrel
Paradise Shelduck
Pied Oystercatcher
Pied Shag
Pied Stilt
Pukeko
Redpoll
Rifleman
Royal Albatross
Royal Spoonbill
Salvins Mollymawk
Silvereye
Skylark
Song Thrush
Sooty Shearwater
Spotted Shag

Stewart Island Shag
Tomtit
Tui
Variable Oystercatcher
Wandering Albatross
Weka
Westland Petrel
White Chinned Petrel
White Faced Heron
White Fronted Tern
Yellow Eyed Penguin
Yellowhammer

TOTAL = 60
BIRD COUNT = 673

Vietnam

Black Capped Kingfisher
Blue Rock Thrush
Burmese Shrike
Chinese Pond Heron
Common Kingfisher
Daurian Redstart
Grey Wagtail
Japanese White Eye
Javan Pond Heron
Olive Backed Sunbird
Pied Fantail
Pied Kingfisher
Stripe Throated Bulbul
Yellow Wattled Lapwing

TOTAL = 14
BIRD TOTAL = 687

Thailand

Ashy Drongo
Ashy Minivet

Asian Fairy Bluebird
Asian Glossy Starling
Asian Openbill
Black Winged Stilt
Blue Eared Kingfisher
Brahminy Kite
Brown Prinia
Brown Winged Kingfisher
Buff Vented Bulbul
Chestnut Bellied Malkoha
Chestnut Flanked White Eye
Common Redshank
Common Sandpiper
Common Tailorbird
Dusky Crag Martin
Eastern Marsh Harrier
Eurasian Curlew
Great Crested Tern
Great Knot
Greater Coucal
Green Bee-Eater
Greenshank
Grey Capped Pygmy Woodpecker
Greater Racket Tailed Drongo
Hoopoe
Indian Roller
Large Billed Crow
Lesser Sand Plover
Little Cormorant
Little Grebe
Long Tailed Shrike
Nordmann's Greenshank
Ochraceous Bulbul
Oriental Magpie Robin
Oriental Skylark
Oriental Turtle Dove
Paddyfield Pipit
Pied Bushchat
Plaintive Cuckoo
Red Avadavit
Red Collared Dove

Red Throated Barbet
Red Wattled Lapwing
Rufous Tailed Tailorbird
Scaly Breasted Munia
Scarlet Backed Flowerpecker
Spotted Dove
Spotted Redshank
Striated Swallow
Terek Sandpiper
White Breasted Waterhen
White Chested Babbler
White Vented Mynah

TOTAL = 55
BIRD COUNT = 742

India

Alexandrine Parakeet
Ashy Bulbul
Ashy Prinia
Asian Barred Owlet
Asian Brown Flycatcher
Asian Koel
Asian Pied Starling
Banded Bay Cuckoo
Bank Myna
Bar Headed Goose
Barwinged Flycatcher Shrike
Bay Backed Shrike
Black Bittern
Black Bulbul
Black Chinned Babbler
Black Crested Bulbul
Black Eagle
Black Headed Gull
Black Headed Ibis
Black Headed Munia
Black Headed Oriole
Black Ibis
Black Lored Tit

Black Naped Monarch
Black Redstart
Black Rumped Flameback
Black Shouldered Kite
Black Stork
Black Tailed Godwit
Blossom Headed Parakeet
Blue Breasted Quail
Blue Whistling Thrush
Blyth's Leaf Warbler
Brahminy Starling
Bronze Winged Jacana
Brook's Leaf Warbler
Brown Dipper
Brown Fish Owl
Brown Headed Barbet
Brown Headed Gull
Changeable Hawk Eagle
Chestnut Bellied Nuthatch
Chestnut Eared Bunting
Chestnut Sh. Petronia
Chestnut Tailed Starling
Chiffchaff
Citrine Wagtail
Clamorous Reed Warbler
Collared Scops Owl
Comb Duck
Common Babbler
Common Coot
Common Crane
Common Hawk Cuckoo
Common Iora
Common Kestrel
Common Merganser
Common Rosefinch
Common Snipe
Common Stonechat
Common Woodshrike
Coppersmith Barbet
Crested Bunting
Crested Kingfisher

Crested Serpent Eagle
Crested Tree Swift
Crimson Sunbird
Dark Throated Thrush
Dusky Eagle Owl
Dusky Warbler
Egyptian Vulture
Emerald Dove
Eurasian Crag Martin
Eurasian Spoonbill
Eurasian Thick Knee
Eurasian Wigeon
Fulvous Breasted Woodpecker
Garganey
Golden Fronted Leafbird
Golden Oriole
Graceful Prinia
Great Tit
Great White Pelican
Greater Coucal
Greater Flameback
Greater Spotted Eagle
Green Sandpiper
Greenish Warbler
Grey Breasted Prinia
Grey Bushchat
Grey Francolin
Grey Headed Canary Flycatcher
Grey Hornbill
Grey Winged Blackbird
Gull Billed Tern
Himalayan Bulbul
Himalayan Flameback
House Martin
Hume's Leaf Warbler
Imperial Eagle
Indian Cormorant
Indian Nightjar
Indian Peafowl
Indian Pond Heron
Indian Robin

Indian Silverbill
Indian Treepie
Intermediate Egret
Jungle Babbler
Jungle Owlet
Jungle Prinia
Kalij Pheasant
Large Cuckooshrike
Large Grey Babbler
Laughing Dove
Lesser Adjutant
Lesser Coucal
Lesser Spotted Eagle
Lesser Whistling Duck
Lesser Whitethroat
Lesser Yellownape
Little Grebe
Little Pied Flycatcher
Little Ringed Plover
Little Stint
Long Billed Vulture
Malabar Pied Wagtail
Marsh Sandpiper
Olive Backed Pipit
Orange Headed Thrush
Oriental Honey Buzzard
Oriental White Eye
Painted Stork
Pale Billed Flowerpecker
Plain Martin
Plain Prinia
Plum Headed Parakeet
Plumbeous Water Redstart
Pochard
Puff Throated Babbler
Purple Heron
Purple Sunbird
Red Breasted Flycatcher
Red Breasted Parakeet
Red Headed Vulture
Red Junglefowl

Red Necked Falcon
Red Rumped Swallow
Red Spotted Bluethroat
Red Vented Bulbul
River Lapwing
River Tern
Rock Bush Quail
Rock Dove
Rose Coloured Starling
Rose Ringed Parakeet
Ruddy Shelduck
Ruff
Rufous Bellied Niltava
Rufous Gorgetted Flycatcher
Rufous Tailed Lark
Rufous Treepie
Rufous Woodpecker
Rusty Tailed Flycatcher
Sarus Crane
Scaly Bellied Woodpecker
Scaly Thrush
Scarlet Minivet
Shikra
Siberian Rubythroat
Slaty Blue Flycatcher
Small Minivet
Small Niltava
Small Pratincole
Southern Grey Shrike
Spangled Drongo
Spot Billed Duck
Spotted Flycatcher
Spotted Owlet
Steppe Eagle
Streak Throated Swallow
Tawny Eagle
Temminck's Stint
Tickells Flycatcher
Tree Pipit
Tufted Duck
Ultramarine Flycatcher

Variable Wheatear
Velvet Fronted Nuthatch
Verditer Flycatcher
Wallcreeper
White Bellied Drongo
White Browed Fantail
White Browed Wagtail
White Capped Bunting
White Capped Water Redstart
White Eared Bulbul
White Eyed Buzzard
White Naped Woodpecker
White Rumped Shama
White Rumped Vulture
White Tailed Lapwing
White Tailed Rubythroat
White Throated Kingfisher
White Wagtail
Wire Tailed Swallow
Wood Sandpiper
Woolly Necked Stork
Yellow Bellied Fantail
Yellow Billed Babbler
Yellow Crowned Woodpecker
Yellow Footed Green Pigeon
Yellow Legged Gull

TOTAL = 213
BIRD COUNT = 955

South Africa

African Black Oystercatcher
African Long Tailed Shrike
African Scops Owl
Arrow Marked Babbler
Bateleur
Black Cuckoo Shrike
Black Headed Oriole
Blacksmith Plover
Blue Waxbill

253

Bokmakierie
Brown Throated Martin
Burchell's Glossy Starling
Cape Bulbul
Cape Bunting
Cape Cormorant
Cape Glossy Starling
Cape Griffon Vulture
Cape Sparrow
Cape Turtle Dove
Cape Wagtail
Cape White Eye
Common Tern
Crested Francolin
Crimson Breasted Shrike
Crowned Plover
Egyptian Goose
Emerald Spotted Dove
European Bee-Eater
European Roller
Fiery Necked Nightjar
Fiscal Shrike
Fork Tailed Drongo
Giant Eagle Owl
Goliath Heron
Grey Lourie
Groundscraper Thrush
Hadeba Ibis
Hamerkop
Hartlaub's Gull
Helmeted Guineafowl
Hooded Vulture
Jackass Penguin
Lesser Grey Shrike
Lesser Striped Swallow
Lilac Breasted Roller
Marabou Stork
Marico Flycatcher
Mountain Chat
Natal Francolin
Olive Thrush

Ostrich
Paradise Wydah
Pearl Spotted Owl
Puffback
Red Backed Shrike
Red Billed Oxpecker
Red Billed Wood Hoopoe
Red Collared Widow
Red Faced Mousebird
Red Winged Starling
Reed Cormorant
Rock Martin
Rock Pigeon
Rufous Naped Lark
Sacred Ibis
Secretary Bird
Southern Greyheaded Sparrow
Southern Yellow Billed Hornbill
Spotted Dickop
Spotted Eagle Owl
Striped Cuckoo
Swainson's Francolin
Swift Tern
White Backed Vulture
White Necked Raven
White Rumped Swift
Woodland Kingfisher

TOTAL = 77
FINAL BIRD COUNT = 1032

TravellersEye Club Membership

Each month we receive hundreds of enquiries from people who've read our books or entered our competitions. All of these people have one thing in common: aching to achieve something extraordinary, outside the bounds of our everyday lives. Not everyone can undertake the more extreme challenges, but we all value learning about other people's experiences.

Membership is free because we want to unite people of similar interests. Via our website, members will be able to liaise with each other about everything from the kit they've taken to the places they've been to and the things they've done. Our authors will also be available to answer any of your questions if you're planning a trip or if you simply have a question about their books.

As well as regularly updating members with news about our forthcoming titles, we will also offer you the following benefits:

Free entry to author talks/signings
Direct author correspondence
Discounts off new and past titles
Free entry to TravellersEye events
Discounts on a variety of travel products and services

To register your membership, simply write or e-mail us giving your name and address (postal and e-mail). Our contact details are:

TravellersEye Limited
51A Boscombe Road
London W12 9HT
United Kingdom

Tel: (44) 20 8743 3276
Fax: (44) 20 8743 3276
E-mail: books@travellerseye.com
Website: www.travellerseye.com

About TravellersEye

I believe the more you put into life, the more you get out of it. However, at times I have been disillusioned and felt like giving up on a goal because I have been made to feel that an ordinary person like me could never achieve my dreams.

The world is absolutely huge and out there for the talking. There has never been more opportunity for people like you and me to have dreams and fulfil them.

I have met many people who have achieved extraordinary things, and these people have helped inspire and motivate me to try and live my life to the fullest.

TravellersEye publishes books about people who have done just this and we hope that their stories will encourage other people to live their dream.

When setting up TravellersEye I was given two pieces of advice. The first was that there are only two things I ever need to know: your are never going to know everything and neither is anyone else. The second was that there are only two things I ever need to do in life: never give up and don't forget rule one.

Nelson Mandela said in his presidential acceptance speech: 'Our deepest fear is not that we are inadequate. Our deepest fear is that we are powerful beyond our measure . . . as we let our own light shine, we unconsciously give other people permission to do the same.'

We want people to shine their light and share it with others in the hope that it may encourage them to do the same.

Dan Hiscocks
Managing Director
TravellersEye Limited

Travels in Outback Australia: Beyond the Black Stump

Andrew Stevenson

A seasoned traveller and travel writer, Andrew Stevenson is unafraid of the unconventional. Whilst most people visiting Australia tread the well-worn path from the Sydney Opera House and up the East Coast via Ayers Rock, Andrew disappeared into the Australian Outback in search of the original Australians – the Aborigines.

'If you want to meet them nowadays, you've got to go beyond the black stump!' he was told.

Going where few have gone before, Andrew delves into the Outback without fear. Drinking in bars with people even the locals avoid, asking questions that we all want to hear the answers to.

Written with humour and compassion, his powers of observation and enquiring mind draw out a frankness that is sometimes shocking but something from which we all can learn. *Travels in Outback Australia: Beyond the Black Stump* is no ordinary tale of an intrepid traveller, it is an extraordinary account of an Australia that we have not seen before.

ISBN: 1 903070 14 7
Price: £9.99

Grey Paes and Bacon

Bob Bibby

Paying no heed to his gammy left knee, Wolverhampton-bred writer Bob Bibby set out to discover the Black Country. He vowed not to return until he'd found out why the dialect is so mellifluous, where black country 'osses come from and what Lenny Henry was like as a schoolboy. Anxious about marching off into the unknown alone, he enlists ex-Scaffold member and Tiswas front man John Gorman as companion and bodyguard.

What ensues is a journey that is in turn entertaining, informative and wickedly irreverent. Join our unlikely heroes as they stamp on the stereotypes, sample the gastronomic delights, and unravel the mysteries, both past and present, of the Black Country.

ISBN: 1 903070 06 6
Price: £7.99

Dancing With Sabrina: A walk from source to sea of the River Severn

Bob Bibby

Following his last best-selling book about a walk around the canals of the Black Country, Bob dons his boots again to explore where the river that runs past his doorstep starts and ends.

Sabrina was the secret love child of the ancient British King Locraine, and the name that the Romans gave to the longest river in England, now called the Severn.

Dancing with Sabrina takes us on a fantastic journey from its source in the Welsh mountains to the Bristol Channel where she meets the sea.

Visiting the towns that live on her banks, Bob whisks us back to discover the past and meet the villains, heroes and madmen that she has known along the way. He samples the modern day with his refreshingly simple needs and throws the light on where to go and not to go, to eat, drink and visit. This book will make anyone from the area or visiting feel lucky to be part of it and make them laugh as they learn about the secrets it holds.

ISBN: 1 903070 24 4
Price: £9.99

Cry From The Highest Mountain

Tess Burrows

'This is an enthralling journey of courage and endurance.'

Joanna Lumley

Their goal was to climb to the point furthest from the centre of the Earth, some 2150 meters higher than the summit of Everest.

Their mission was to promote Earth Peace by highlighting Tibet and His Holiness the Dalai Lama's ideals of peace, harmony and justice as an arrow of light for the new millennium.

But for Tess it became a struggle of body and mind as, in the ultimate act of symbolism, she was compelled to climb to the highest point within herself.

'I am pleased to offer my prayers to "Climb For Tibet" and all that it touches.'

His Holiness the Dalai Lama

ISBN: 1 903070 12 0
Price: £7.99

OTHER TITLES FROM TRAVELLERSEYE

Jasmine and Arnica
Nicola Naylor
Young blind woman travels alone around India.
ISBN: 1 903070 10 4

Triumph Round the World
Robbie Marchall
He gave up his world for the freedom of the road.
ISBN: 1 903070 08 2

Jungle Janes
Peter Burden
Twelve middle-aged women take on the Borneo Jungle. Seen on Channel 4.
ISBN: 1 903070 05 8

Desert Governess
Phyllis Ellis
An inside view of the Saudi Royal Family
ISBN: 1 903070 01 5

Travels with my Daughter
Niema Ash
Forget convention, follow your instincts.
ISBN: 1 903070 04 X

Fever Trees of Borneo
Mark Eveleigh
A daring expedition through uncharted jungle.
ISBN: 0 953057 56 9

What for Chop Today?
Gail Haddock
Experiences of VSO in Sierra Leone.
ISBN: 1 903070 07 4

Discovery Road
Tim Garrett and Andy Brown
Their mission was to mountain bike around the world.
ISBN: 0 953057 53 4

Riding with Ghosts: South of the Border
Gwen Maka
Second part of Gwen's epic cycle trip across the Americas.
ISBN: 1 903070 09 0

Fridgid Women
Sue and Victoria Riches
The first all-female expedition to the North Pole.
ISBN: 0 953057 52 6

Jungle Beat
Roy Follows
Fighting Terrorists in Malaya
ISBN: 0 953057 57 7

Slow Winter
Alex Hickman
A personal quest against the backdrop
of the war-torn Balkans.
ISBN: 0 953057 58 5

Riding with Ghosts
Gwen Maka
One woman's solo cycle ride from
Seattle to Mexico.
ISBN: 1 903070 00 7

Tea for Two
Polly Benge
She cycled around India to test her love.
ISBN: 0 953057 59 3

Touching Tibet
Niema Ash
One of the first Westerners to enter
Tibet.
ISBN: 0 953057 550

Traveller's Tales from Heaven and
Hell

More Traveller's Tales from Heaven
and Hell
Past winners of our competitions.
ISBN: 0 953057 51 8
and 1 903070 02 3

A Trail of Visions: Route 1 – India,
Sri Lanka, Thailand, Sumatra
A Trail of Visions: Route 2 – Peru,
Bolivia, Columvia, Ecuador
Vicki Couchman
A stunning photographic essay.
ISBN: 1 871349 33 8
and 0 953057 50 X